Eisenhower at the Dawn of the Space Age

Eisenhower at the Dawn of the Space Age

Sputnik, Rockets, and Helping Hands

Mark Shanahan

LEXINGTON BOOKS
Lanham • Boulder • New York • London

Published by Lexington Books
An imprint of The Rowman & Littlefield Publishing Group, Inc.
4501 Forbes Boulevard, Suite 200, Lanham, Maryland 20706
www.rowman.com

Unit A, Whitacre Mews, 26-34 Stannary Street, London SE11 4AB

British Library Cataloguing in Publication Information Available

Library of Congress Cataloging-in-Publication Data

Names: Shanahan, Mark, author.
Title: Eisenhower at the dawn of the Space Age : Sputnik, rockets, and helping hands / Mark Shanahan.
Description: Lanham : Lexington Books, [2016] | Includes bibliographical references and index.
Identifiers: LCCN 2016041300 (print) | LCCN 2016042803 (ebook) | ISBN 9781498528146 (cloth : alkaline paper) | ISBN 9781498528153 (Electronic)
Subjects: LCSH: Astronautics and state--United States--History--20th century. | Outer space--Exploration--Government policy--United States--History--20th century. | Eisenhower, Dwight D. (Dwight David), 1890-1969--Political and social views. | United States--Relations--Soviet Union. | Soviet Union--Relations--United States. | Space race--United States--History--20th century.
Classification: LCC TL789.8.U5 S434 2016 (print) | LCC TL789.8.U5 (ebook) | DDC 629.40973--dc23
LC record available at https://lccn.loc.gov/2016041300

∞™ The paper used in this publication meets the minimum requirements of American National Standard for Information Sciences Permanence of Paper for Printed Library Materials, ANSI/NISO Z39.48-1992.

Printed in the United States of America

To Jac for patience way beyond the call of duty

Contents

Introduction

When I was five years old, man landed on the moon. It was the most truly amazing feat I could imagine. Through each of the Apollo missions I'd stand in my back yard in London, England, and stare at moon's surface 8 x 30 binoculars in hand, trying desperately to see some sign of movement; some sign that man had made it to that cold world. But then, by the time I was eight, it was all over. However, the nugget of the space story stuck in my mind and persisted across more than forty years. But was what actually happened quite the same as the story propagated in hundreds of books and countless repetitions on television of a sequence that became the accepted norm? It started with Sputnik, a soviet satellite success that left America in panic. Hampered by an aging "do-nothing" president, America failed to react. Nothing changed until the presidency didn't just change hands, but changed generations. The young, good-looking and talismanic John F Kennedy entered the White House and, at least in the TV re-telling, miracles happened. He stood in front of the two Houses of Congress and told America's political elite that he pledged to send a man to the moon before this decade was out. And despite the tragic events in Dallas in November 1963, it happened. Kennedy's legacy was fulfilled and man walked on the moon.

It was with that in mind that I set out on a research project in 2010. From some research at NASA's History Office in Washington, DC and at the Kennedy Presidential Library at Columbia Point, I already knew that the reality of Kennedy's interest in space was different from the rhetoric and I had a hypothesis that suggested the real reason for Kennedy's drive for space was media pressure. I was looking to prove that causative impact. To do so, I needed the context for Kennedy's actions. I needed to go back to the previous president. I needed to understand what role, if any, the lackluster Eisenhower played in letting the Soviet Union win all the early battles in the space race. My return visit to delve into the NASA historical collections coupled with my first visit to the Eisenhower Presidential Library in Abilene, Kansas, was a sobering one. Within a couple of days, I had shot down my initial hypothesis. There was no causative link between the media and Kennedy's actions. A couple of weeks later, after many hours getting to grips with the Eisenhower papers, I had changed research direction totally. Ike emerged as the singularly more interesting character: the man who had a plan for space that was built on national security and bounded by fiscal pragmatism. Eisenhower did not react to Sputnik in the way that seems ingrained in the popular memory. The evidence simply didn't back that. Instead I found a long-game president; a planner whose actual activities around space operated on a parallel path to the oft-told story. I found a leader whose presidency had been underrated—and where his handling of Sputnik had moved from misreporting to footnotes. He deserved better.

So, as the sixtieth anniversary of the Sputnik 1 launch approaches, are we correct in our reading of the Eisenhower presidency?

> Ike was a mid-western middle-of-the-road patriot, a common man with a winning smile who read little, was uninformed about trends in intellectual and artistic life, and was prone to giving folksy advice. A diplomat labelled him "the nation's number one Boy Scout." But those who were closer to Ike have presented another side. When provoked, the genial general could show a fiery temper and release a stream of scalding profanity.[1]

This is how George Tindall and David Shi sum up President Dwight D Eisenhower in *America, A Narrative History*. In the same paragraph, however, they note that "Eisenhower was, in fact, an effective leader. . . . One student of Eisenhower's leadership techniques has spoken of "a hidden-hand presidency" in which Ike deliberately cultivated a public image of passivity to hide his active involvement in policy decisions."[2] It is rather a mixed message—and Tindall and Shi give no indication of their judgment on the veracity of the argument of that "One student."

So, do we make assumptions about Eisenhower? Are we correct in the way we read his presidency not just through his character, but through his actions, and in particular, his actions in the wake of the launch of Sputnik? Hugh Brogan, in *The Penguin History of the USA*, still a standard work for students, neatly expressed the orthodox view:

> The Eisenhower years were, in general, ones of comfortable lethargy. When the Soviet Union put the first satellite into space in 1957, the shock to American vanity was almost unbearable; the cry went up that something was badly wrong with American society, American science, American education; it would take several years for the speed with which the lapse was made good to wipe out this impression.[3]

Eisenhower's presidency is skated over in just a few pages while the supposed Sputnik crisis is dismissed in a third of a paragraph. It is true that Brogan's work was first written in 1985, but this assessment of "The Eisenhower Years" remained unchanged through two editions and a number of reprints. David Reynolds, in another more recent narrative history regularly used by students, is rather less pejorative in his assessment of Eisenhower and Sputnik, but still deals with the issue very briefly. In *America, Empire of Liberty*, he wrote about the wake of the two Sputnik satellite launches.

> Criticism of Eisenhower had already been mounting: there were reports of heart trouble, and he seemed to be spending too much time on the golf course. Now, America's leader looked dangerously out of touch . . . Sputnik was a huge shock to America's national pride: Eisenhower immediately established the National Aeronautics and Space Agency [sic—it should be Administration] to coordinate America's space race.[4]

Reynolds now repeats the traditional narrative for Eisenhower and Sputnik: the weak president reacts to a Soviet triumph. It is a narrative that has been accepted since the first generation of scholars began to study the Eisenhower presidency and it has become a totem both for Eisenhower scholars and for anyone interested in early space exploration. Even those directly involved in those early US endeavors present an unquestioning adherence to the expected narrative. Neil Armstrong summed this up:

> Sputnik did change our world . . . President Eisenhower was saying something like: "What's the worry? It's just one small ball." But I'm sure that was a façade behind which he had substantial concerns, because if they put something into orbit, they could put a nuclear weapon on a target in the United States.[5]

Eisenhower at the Dawn of the Space Age questions that narrative, adding to the ongoing revision of Eisenhower's actions on space policy and in so doing, re-interpreting the view that Eisenhower's second term in particular was one of lame-duck inaction.

Sixty years on from the "Sputnik Fall" of 1957, The time is ripe for reassessing Eisenhower's presidency. In recent years, a slew of biographies by Evan Thomas, Jim Newton, Paul Johnson, and Jean Edward Smith have all added to the rehabilitation of a president who, for too long, was regarded as a passive political player. He had been seen as content to inhabit the golf course rather than the White House; happier on the farm in the company of his Kansas friends than on the international stage battling for the preeminence of the free world. But while there has been a renaissance of interest in Eisenhower among presidential biographers, there has been too little scholarly advance on the revisionist stance set by Fred Greenstein—Tindall and Shi's "One student"—in his 1982 book, *The Hidden-Hand Presidency*.[6] That picture is changing somewhat, not least through the work of David Nichols, whose detailed studies of Eisenhower's actions around civil rights, the Suez Canal, and Ike's relationship with Joe McCarthy are throwing a new light on Eisenhower as president—not quite the all-knowing political master as envisioned by Greenstein, but also far from the benign, grinning, indecisive and ineffective non-leader portrayed by the first generation of scholars to pass judgment once he had left political office in 1961. This book builds on the work of scholars such as Nichols and provides a new lens on one particular aspect of Eisenhower's presidency: his development of a comprehensive policy for the United States' first endeavors in outer space.

Today's emerging view of "Ike" as he was commonly known, is of a long-term strategist; a smart thinker who, unlike previous interpretations, was a decisive decision maker. But this view needs more extensive and scholarly drawing-out. While biographers from Stephen Ambrose to Jean Edward Smith engage lightly on Eisenhower's talents through a broad sweep approach and a determination to tell a good story well, there is insufficient space in a single volume to delve deeply into the detail of a two-term presidency, not least when the "back story" of Eisenhower's military career either dominates proceedings or at least casts a long shadow over his presidency. Speaking at the "Ike Reconsidered" Conference in New York in 2013, Nichols summed up the current scholarly debate as "midstream" and provided a pithy summary of Eisenhower as president: "The real Dwight Eisenhower was a crafty, complicated man. Agree with him or not, he was formidable."[7] To understand just how formidable he was, it is necessary to unpick specific areas for deeper study—as Nichols has done with Suez, civil rights, and, most recently, McCarthyism.

This book adds to the investigation of specific facets of the Eisenhower presidency and contributes to the knowledge of his presidency by focusing on the development of the United States' first space policy. It returns to the question: Have historians read Sputnik wrong? Indeed, should we instead consider why Eisenhower *did not* react to Sputnik? In so doing, it raises further questions. For instance, it asks: To what extent was Eisenhower actually an "active" president in his second term and how did he demonstrate his activism, particularly in dealing with Congress in the wake of the Sputnik launches? In seeking to answer this, this book asks whether historians should revisit Greenstein's revisionism to build from his excellent "straw man" of "Hidden Hand" presidential action. By using these questions to lead an investigation into archive materials—especially primary source documentation, supported by memoir and oral histories—and by reinterpreting existing secondary sources, this study shows Eisenhower already had plans for both military uses of space and for scientific satellites well before 1957 and thus had no need to panic over Sputnik.[8] He was operating on a parallel path to the knee-jerk political calls for ever greater weaponry in the wake of the Sputnik launches. Rather than react to Soviet action, he was driven by the need to balance his overriding concern to ensure national security with his ingrained economic conservatism. By an in-depth analysis of the development of Eisenhower's space policy from 1953, *Eisenhower at the Dawn of the Space Age* also challenges the prevailing scholarly view of the path of the National Aeronautics and Space Act, traditionally seen as a congressional achievement, led by Senate majority leader, Lyndon Johnson. This study reinterprets the evidence, looking particularly at the role of the executive, and questioning who was using whom (and who benefited most) from the executive/legislative interaction.

This book remains located within Fred Greenstein's "Hidden Hand" view of Eisenhower. But in order for the "Hidden Hand" presidency to be effective, Eisenhower required "Helping Hands." Those figures are at the heart of this work as I ask what role "operational officers" occupied in Eisenhower's decision making. My focus is on identifying who they were and analyzing their influence on the policy process—either as definers or appliers of policy. I discuss what kind of personalities Eisenhower looked for, and how he was able to achieve policy goals—often at odds with either a hostile Congress, Pentagon, or both—by giving the support his advisers needed to reach key decisions, and then backing them in applying those decisions. This is not a specific study of Eisenhower's military missile policy or of his intelligence community's strategy and tactics in dealing with the Soviet threat. It does, however show both of these important aspects of his presidential control, if only where they had significant influence on key elements of the development of Eisenhower's space policy.

In choosing to dissect Eisenhower's decision making by reviewing the development of his space policy, this study will revisit ground first covered in the 1990s by Robert Divine in his monograph *The Sputnik Challenge,* and the more journalistic chronicle *Sputnik: the Shock of the Century,* written in 2000 by Paul Dickson.[9] However, Divine, Dickson and even Yanek Mieczkowski's 2013 work, *Eisenhower's Sputnik Moment* all follow one orthodox line that paints Eisenhower's actions in regard to outer space as a *reaction* to Sputnik.[10] Meanwhile, much of the wider early space discourse focuses on President Kennedy's 1961 pledge "to send a man to the moon and return him safely to the earth before this decade is out." This views the early space program through a distorting lens; one that throws undue focus on the "triumph" of the moon landing. Eisenhower's initiative in the creation of the infrastructure for the United States' space activities is underplayed. His motives and the process that created a "white" space program for the United States alongside a highly-effective "black" program focused on reconnaissance tend to be discussed only as a prelude to the Kennedy/Johnson program that delivered the first moon landing. Yet this undervalues considerably Eisenhower's contribution to the United States' achievements in space in the first decades of space exploration.

One of the key questions the book engages with is the value to the United States of its space-related efforts. Traditionally, the success of US space endeavors is seen in the prestige gained from the Apollo moon landings. Kennedy gained congressional support for the race to the moon which has been presented by scholars as a proxy battle in the Cold War.[11] However, there has been relatively little critical engagement with the period *before* Kennedy, other than to present Eisenhower as delivering a weak response to Sputnik. This work engages with that pre-Kennedy period and, by drawing together Eisenhower's actions around missiles, satellites, and the separation of military and scientific space activities, presents an account of the United States' success in space based not on Kennedy's pledge to send a man to the moon, but on the immense amount of groundwork achieved under Eisenhower. Prestige and credibility mattered as much to Eisenhower as to any other president, and in a time when new nations were emerging from colonial rule to independence across Asia and Africa, the prestige of the United States was vital. Eisenhower was no different from any other American in wanting these new nations to opt for the "free world" path offered by the United States rather than the opposing ideology led by Khrushchev's Soviet Union.[12] But I question the degree to which he was interested in making outer space activities into an ideological race.

This study joins the discourse on the politics of the early years of space endeavor by distilling Eisenhower's process of moving to a civilian space agency reporting directly to the president. Prior to, but connected to, this process, it analyzes his response to nuclear missile testing and his decisions on the U-2 and CORONA "black" reconnaissance programs. It explains how all these factored into his decision not to seek Cold War confrontation over space exploration. It explores the effect of Eisenhower's preference for unmanned space missions; and desire to separate scientific and military uses of space. In so doing, it demonstrates a significant policy difference from the direct challenge to Khrushchev made by Kennedy which coalesced all NASA's focus into the race for the moon.

While this study is primarily focused on the development of the first US space policy, it contributes to the overall revaluation of Eisenhower underway at present. In touching upon Eisenhower's relationship with Johnson in the Senate, for instance, it considers the strength of the executive, when the orthodox reading of Eisenhower's second terms tends to stress the dominance of the legislative branch. Through the very specific lens of space policy making, I aim to add to the weight of evidence emerging to support the claim that Eisenhower was an effective, activist president.

In summary, as we approach the sixtieth anniversary of the dawn of the Space Age, *Eisenhower at the Dawn of the Space Age* presents a new perspective on Eisenhower's contribution to the United States' early space policy development. Understanding how he acted on missile, rocket, and space issues, from his first term to his final days in the White House, has significant implications for the broader understanding both of the US exploration of space and for Eisenhower's presidency.

NOTES

1. G Tindall and D Shi, *America, A Narrative History*, 7th edition, volume 2 (New York, 2007), p. 1199.
2. Ibid.
3. H Brogan, *The Penguin History of the USA* (London, 1999), p. 612.
4. D Reynolds, *America, Empire of Liberty* (London, 2010), p. 404.
5. Neil Armstrong in JR Hansen, *First Man, the Life of Neil Armstrong* (London and New York, 2005) pp. 168–69.
6. F Greenstein, *The Hidden Hand Presidency* (Baltimore, 1982).
7. D Nichols, closing address at "Ike Reconsidered—Lessons from the Eisenhower Legacy for the 21st Century" Conference held at Hunter College CUNY in association with The Eisenhower Foundation. Sourced from roosevelthouse.hunter.cuny.edu/ike/. Accessed April 4, 2013.
8. A Schlesinger Jr, *The Imperial Presidency* (New York, 2004 edition); R Neustadt's *Presidential Power* (New York, 1960).
9. R Divine, *The Sputnik Challenge* (Oxford, 1993); P Dickson, *Sputnik, The Shock of the Century* (New York, 2001).
10. Y Mieczkowski, *Eisenhower's Sputnik Moment* (Ithaca and London, 2013).
11. M Beschloss, "Kennedy and the Decision Go to the Moon," in R Launius and H McCurdy, eds., *Spaceflight and the Myth of Presidential Leadership* (Urbana, 1997), pp. 51–53.
12. This is well stated by Walter La Feber in *America, Russia and the Cold War, 1945–1996*, 8th Edition (New York, 1997) pp. 192–93 when he says: "The newly emerging nations could view Russia as a people who in 1917 had been generations behind other industrialized nations but who, through harsh regimentation, had assumed first place in the race for control of outer space. They could also interpret the launching as a dramatic swing in the balance of military power towards Moscow."

ONE

The Policy-Making Character

There is a statement, sometimes attributed to Seneca the Younger that says: "Luck is what happens when preparation meets opportunity."[1] Eisenhower is often regarded as a lucky general who was lucky to be president.[2] But he epitomizes this sentiment. Throughout his life, from West Point to the White House, he worked hard to make the most of his abilities; to cultivate the right people and to ensure he was in the best position to benefit when opportunities arose. It served him well through most of his military career as he aligned himself alongside the great figures of the Army's general staff. It served him well at war as he surrounded himself with the best operational military talent who ensured he never had to enter a battle without the knowledge that he had the blend of forces, collateral and planning to enable him to win. And it served him well when as a "non-politician" he emerged to take the presidency in the 1952 election. His style throughout his eight years in the White House was significantly more military than political and offers much to explain the chain of decisions around space policy that he took from 1953 to the weeks before he finally left office in 1961.

If this was simply an "airport paperback," it would jump straight into the "outer space" action: probably starting with the launch of Sputnik. And if it is the supposed reaction to Sputnik that interests you most, you may care to speed-read through to the next chapter. But the aim of this book is to revisit the space policy decisions Eisenhower made within a scholarly framework. Therefore, using some of the more important stops on the historiography trail, it is necessary to define the character and the style of the man who made the key decisions as the United States developed its first space policy.[3] That is the purpose of this chapter. Having painted a very short biographical picture of Eisenhower and the state of the nation he inherited when he entered the White House in 1953, it will assess his current standing in the eyes of historians as well as indicating how this picture has emerged. Following this, the chapter will move on to cast the political scientists' eye over Eisenhower, and also cover the historiography of the early space age. It will conclude with a brief passage on how Eisenhower saw himself. Each section will build the context within which his actions around Sputnik and the development of space policy should be considered.

EISENHOWER IN 1953

The Eisenhower who entered the White House in 1953 was not young. He was sixty-two years old, significantly younger than both Hillary Clinton and Donald Trump, but seemingly older when the average life expectancy for a white male in the United States was 66.5 years.[4] To his opponents, his best years had passed. A life-long career in the military has culminated in his role as Supreme Commander, Europe, during World War II. But to his supporters, this had been simply an essential step on his journey to the White House. Eisenhower was a hero for America, and had run for the presidency at a time when America needed heroism to face the scourge of communism, not least in Korea where the conflict had proven all too bloody and attritional. From graduating from the West Point Class of 1915, he had been in the military for all but two years of his adult life, with those years spent as president of Columbia University from 1948 to 1950. And in the run-up to the presidential election, he had spent two years based in Paris as the Supreme Commander of the newly-formed NATO. Resigning his commission in June 1952, he was nominated as the Republican candidate for the presidency at the GOP Convention in Chicago following a hard-fought ballot that had pitched him against Senator Robert Taft, the Republicans' principal congressional spokesman, and lesser candidates Governor Earl Warren, Minnesota's Harold Stassen, and even General Douglas MacArthur, who had been sacked from his role leading the Allied forces in Korea by President Truman.

Truman's post-1948 term in the White House was running into severe turbulence with the post-World War II economic boom slowing down at home with fears over rising prices, rising taxes, and a rising national debt. Abroad, all Americans saw was stagnation in the Korean War and the increasing power and thus threat of the USSR and China. At home, that chilling of the Cold War translated to a fraught atmosphere across the nation as Senator Joe McCarthy and the House Committee on Un-American Activities ramped up their Red Scare activities. After two decades of Democrats in the White House, the 1952 election provided the best chance in a long while for the Republicans to unseat the incumbents, and Eisenhower, the five-star general, was judged the best man for the job. While never party-political in his army years, Eisenhower considered himself to be aligned "fairly closely with what I call the progressive branch of the Republican party."[5] Having seen off the GOP's conservative wing embodied by Taft at the Chicago Convention, Eisenhower's attack on corruption in Washington, the spread of communism and the war in Korea propelled him to an untroubled victory in the November general election. While the GOP squeezed out a slim House majority, Eisenhower easily saw off the challenge of Democrat Adlai Stevenson with 55 percent of the popular vote and a massive win in the electoral college, with 442 votes to his opponent's 89. The new president had come far from his Midwestern small town roots. The man who moved into 1600 Pennsylvania Avenue was a born organizer; a political general whose logistical mastery had made him such an important figure whether in getting the right forces in the right places to prosecute the key battles of World War II, or after 1945, redefining the Army for the new challenge of the Cold War, or in organizing the United States' allies for the defense against the USSR through NATO. Few would have recognized the young man from a strict Mennonite background who had started his working life at Belle Springs Creamery in Abilene, Kansas, forty-three years before. The supposed non-politician who became president was wholly an Establishment figure, firmly at home in Washington and surrounded by "The Gang": key figures in US commerce and industry.

THE HISTORICAL PERSPECTIVE

Eisenhower's style of presidential decision making has been studied since he was in the White House, beginning with the affectionate but often patronizing coverage he received from leading press columnists while still in office, James "Scotty" Reston of the *New York Times* won a Pulitzer prize for his reporting of the president's heart attack and its implications for government in 1957,[6] and continuing into the first academic studies that emerged during the presidencies of Kennedy and Johnson. Eisenhower was operating in a hostile political environment, with political control of the House of Representatives in the era held by the Republican Party only between 1947 and 1949 and during Eisenhower's first term, between 1953 and 1955, while the Senate was in Democratic hands throughout. Having to work with a Democrat Congress undoubtedly limited Eisenhower's choices in policy making and the strategies to implement them, but also had a significant impact on those who chose to critique those policies. Such early critics as Arthur Schlesinger and Richard Neustadt were at the fulcrum of a history and political science scene that shared many of the traits of the liberal political environment. At the time they were writing, expansive, interventionist presidencies, the very antithesis of Eisenhower's political philosophy, were in vogue. Thus their writing is very much a product of the environment they were operating within. A "Whiggish" conservative such as Eisenhower, was simply insufficiently dramatic to reap the praise of this first generation of post-Eisenhower scholarship.[7]

Of course, Neustadt and Schlesinger were not the first writers to hasten Eisenhower's reputational decline in the 1960s. The orthodox view was established by Emmet Hughes, whose scathing review of Eisenhower, written in 1962, deemed him unsuitable for executive office, preferring golf and bridge to reforming the 1950s Republican Party and driving forward policy around civil rights and social reform.[8] This set the tone for a number of writers of whom Neustadt and Schlesinger's work remain the most prominent and influential. Richard Neustadt wrote off Eisenhower as a do-nothing president in *Presidential Power*.[9] Originally a bureaucrat in Truman's Bureau of the Budget, Neustadt lost his role when the Democrats lost out to Eisenhower in 1952 and he switched to academia. However, he reappeared in the political frontline in 1960 as part of Kennedy's transition team, helping to formulate policy and suggest appointments to the new administration. That role came in the wake of the highly-partisan *Presidential Power, The Politics of Leadership*, a work he revised several times to bring succeeding presidents to account. Neustadt's original thesis rated three mid-century presidents: Roosevelt, Truman, and Eisenhower against three attributes—the formal powers of the president, professional reputation, and popularity. While Eisenhower is seen as popular, and a "hero seeking national unity," Neustadt heaped him with faint praise as a political operator: "Through Eisenhower's first six years his power sense was blunt in almost the degree that FDR's was sharp."[10] Despite the fact that the Whitman files were opened in the 1970s, offering scholars primary source evidence to reassess their view of Eisenhower, it is notable that Neustadt chose not to revise his appraisal of Eisenhower in any of the further editions before his death. But Neustadt did make one important point that is certainly true of Eisenhower. He noted that Eisenhower removed many of what Neustadt called "Rooseveltian assignments"—what he regarded as political "yes men"—from the White House, preferring to surround himself with the subtly-different like-minded advisers from beyond partisan political elites, run almost on military lines.[11] For the purposes of this book, these characters fit into the class of "Helping Hands." Indeed, Eisenhower's White House staff was small and close knit. Stephen Hess observed that it comprised just fifty-six posts, with just eighty-eight staff filling these roles across the eight years of Eisenhower's presidency.[12] Perhaps the key insight is Neustadt's suggestion of "like minded" advisers. Eisenhower's instinct was not radical. It was to appoint advisers who, by conviction, shared his social and economic instincts. They may have proved effective in translating big ideas into workable solutions that did push at Eisenhower's boundaries, but these rarely, if ever, pushed the president into completely new concepts.

One historian close to Neustadt's opinion on Eisenhower was another former Kennedy staffer, Arthur Schlesinger Jr. In *The Imperial Presidency,* Schlesinger defined Eisenhower as an "aggrandizing" president, whose assertion of executive privilege "ushered in the greatest orgy of executive denial in American history."[13] Schlesinger asserted that Eisenhower upset the balance of power in US constitutional politics, with the executive branch accumulating power at the expense of Congress.[14] However, this absorption of power was undercut by Eisenhower's unwillingness to use the power effectively. But *The Imperial Presidency* pivots around Nixon rather than Eisenhower, and Schlesinger was more prone to vent his criticism of Eisenhower through his reviews of the writing of others. In 1983, in *The Ike Age Revisited,* his tone was acid as he reviewed the first wave of Eisenhower revisionism.[15] The review was dismissive of a slew of authors who had the temerity to challenge the orthodoxy. Taken with similarly orthodox views from Sherman Adams (removed from his role as chief of staff by Eisenhower—albeit reluctantly—in 1958 for having received gifts in return for influence) and Arthur Larson, briefly a speechwriter for Eisenhower for a year from 1957–1958, and something of a big-government Republican, Schlesinger presented Eisenhower as: "a man of force, dignity and restraint who did not always understand and control what was going on, was buffeted by events, and was capable of misjudgment and error." But while damning Eisenhower with faint praise here, he reserved his stiletto for the final sentence: "Yet we were wrong to have underestimated Eisenhower's astuteness in self-presentation—the best evidence of which perhaps lies in his capacity to take in even intelligent historians."[16]

Yet Schlesinger largely railed against the prevailing mood. Robert Ferrell's *Eisenhower Diaries,* sought to unlock the private thoughts behind the public actions of Eisenhower. This was clearly difficult for Ferrell as Eisenhower was not a prolific diarist, particularly in his presidential years. Schlesinger's chief criticism was that Ferrell's work is incomplete.[17]

In one of the most recent in-depth historical studies of the president, *Eisenhower in War and Peace,*[18] Jean Edward Smith defines Eisenhower's presidential style well. However, he does so not particularly when looking at the presidency, but by discussing Eisenhower's attributes first as a US Army staff officer, and then as supreme commander, in charge of Allied forces in the European Theater in World War II. As a staff officer, Eisenhower had learned the art of strategic command from the very best leaders in the Army. Over twenty years he had served under Fox Conner and George Mosely, described by Smith as the "intellectual kingpins of the interwar army."[19] He had worked for General John Pershing, commander in chief of the World War I American Expeditionary Force and later US Army chief of staff, and Douglas MacArthur, who was Eisenhower's commanding officer for a further seven years. Finally, he reported to General George Marshall who set him to planning how the United States was to hold on to the Philippines in the face of Japanese expansionism, before dispatching Eisenhower to Europe.

Eisenhower was not a fighting general. He had none of the innate battle skills of a Patton or Montgomery, but was peerless in getting the right operational officers around him; getting strong personalities with very different outlooks to work together, and commanding the total theater, not just the individual thrusts of frontline forces. He became exceptionally skilled at planning and regarded the necessity to plan as paramount. Eisenhower, as exemplified by this time in Washington early in World War II was a meticulous planner himself.[20] Indeed, in *Six Crises,* his former vice president, Richard Nixon, quoted Eisenhower as saying: "In preparing for battle, I have always found that plans are useless but planning is indispensable."[21] Essentially, this meant that the art of planning and the insight and understanding it would deliver about any situation was invaluable. But Eisenhower was perfectly aware that no scenario discussed in a planning meeting would ever appear in quite the imagined way in the reality of battle. Plans enabled everyone to prepare to the same level, but what really mattered was the ability to use the insight gained to meet the realities on the battleground. This was the style of leadership he took into the White House. Once there, he operated a more flexible decision-making process than the strict chain of command of the military in so far as he was willing to be challenged by experts, but he was always the final decision maker.

From each of his mentors, and Marshall in particular, Eisenhower had learned the value of the big picture. He set the grand plan, and expected his subordinates to deliver the necessary outcomes. General Bedell Smith said of Marshall and Eisenhower: "Both knew how to delegate. When they assigned a task, they stepped aside. Subordinates were free to follow whatever course they wished to get the job done, it was the old Army at its best."[22] Eisenhower carried his military experience into the presidency, continually assessing the strategic situation, setting tasks for his White House staff and executive appointees and expecting them to fulfil his wishes. Unlike his public persona which could obfuscate when necessary rather than clarify, his White House instructions to subordinates were generally clear and unambiguous.[23] Lucius Clay, who had served under Eisenhower in World War II, described his former commander's style: a style that was relentlessly demanding. "General Eisenhower was not the easiest person in the world to work for. He would give you a job, and when you completed it, he would give you another. The more you did, the more he asked. And if you did not measure up, you were gone. He had no tolerance for failure."[24] In essence, it is arguable that Eisenhower was the first modern president, one who treated his role as being the nation's chief executive officer.[25] It meant setting the tone and direction of policy, but expecting others to do the "heavy lifting'" of operational implementation. However, Eisenhower was always accountable to the American public, and set a very clear expectation of what he required.

Smith, with *Eisenhower in War and Peace*, places his focus on Eisenhower's military career rather than his time in the White House. He emphasized that Eisenhower's great skills came in managing difficult people, especially Churchill, De Gaulle, Patton, and Montgomery, and in taking a high command role that left battlefield operations in the hands of his expert operational subordinates. Indeed, he noted that when Eisenhower attempted to lead at an operational level, such as in the early stages of Operation Torch (the Allied invasion of the western part of North Africa) his skills did not match the needs of the forces on the ground. He was cautious in the field, and too willing to listen to the views of others. The campaign didn't achieve its aims until Eisenhower had handed over the reins of operational command to the fighting specialists.[26]

When he discussed Eisenhower's presidency of Columbia University, Smith noted that Eisenhower regarded the deans and department chairmen as the most important people on campus—the equivalents of army corps and division commanders. As a consequence, he devoted his time to dealing with them. However, this was a total misreading of the situation since, according to Smith: "it is the scholars who determine policy. They are the university."[27] This misreading, which went distinctly against the grain with such scholars as C. Wright Mills, a Columbia University professor (and more from him a little later), marked Eisenhower as an outsider in the politics of the university—a trait repeated in the White House where he chose to mix socially with, and take the advice of, "The Gang," rather than operate at close quarters with the more overtly partisan politicians of the Republican Party, including his vice president, Richard Nixon. "The Gang," as described by Smith were a group of men who were: "Rich, Republican and devoted to golf and bridge." The group, who became life-long friends to Eisenhower, included William Robinson, publisher of the *Herald Tribune*; investment banker (and founder of the Augusta National Golf Club) Cliff Roberts; Coca-Cola Chairman, Bob Woodruff; Ellis Slater, president of Frankfort Distilleries; and Pete Jones, another investment banker.[28] In the 1950s, they represented a particular wing of the Republican Party—the "Wall Street" wing as opposed to the "Main Street" wing inhabited by most congressional Republicans. None of "The Gang" was backward in offering President Eisenhower policy advice, and he used them throughout his two terms as a sounding board for policy creation.[29] Indeed, in filling significant cabinet positions, Eisenhower looked to America's corporate world rather than its political world for men who would bring commercial rigor and good business order to organizing major departments. Defense Secretary Charles Wilson was the president of General Motors, while his successor, Neil McElroy, was president of Procter & Gamble.

Finally, in assessing Eisenhower's 1952 presidential campaign, Smith provided an insight for Eisenhower's preferred modus operandi: one that had not changed since military service and would not change significantly throughout eight years in the White House. He noted that the president assumed control of the broad outlines of the campaign but left the details to others. "He not only set the tone, but made the major decisions." This is an apt description of the president: strategic, logical, organized but not so much prepared to entrust operational duties to others, but adamant that this was necessary for the successful completion of the campaign process. In that respect, the presidential campaign paralleled his earlier military successes.

Evan Thomas, another recent Eisenhower biographer, looked at the hinge points of Eisenhower's presidency through a rather more reductionist lens. Everything he wrote had to fit the theme of his title: *Ike's Bluff*.[30] For Thomas, Eisenhower was a genius at poker and bridge, and used the poker player's guile to keep the United States out of war with the USSR and China over a period when it would have been far easier to fight with weapons rather than the threat to use retaliatory powers that may have appeared more powerful than they actually were. Thomas's portrayal of Eisenhower is more nuanced than Jeffery Frank's in his recent contribution, *Ike and Dick*,[31] but by making every foreign policy decision that Eisenhower considers into a matter of sleight of hand to wrong-foot the Soviets, the picture that emerges of Eisenhower is too often one-dimensional, and his White House policy making too often single-track. But his observation of Eisenhower's personality is succinct:

> Eisenhower could be moody and temperamental. His was not the confidence of the weak, the arrogance of the vain and needy. Rather he had that kind of confidence that allowed him to be humble. He was willing to appear slower and sweeter than he really was in order to get other people to do his bidding. . . . Eisenhower knew that he was strong and that he could see around corners. He did not feel the need to constantly prove his strength . . .
>
> Eisenhower was a great peacekeeper in a dangerous era . . . [he] understood the nature of war better than anyone else and . . . had the patience and wisdom, as well as the cunning and guile, to keep the peace.[32]

In essence, Thomas presents a particular style of poker player: a president who definitely played his cards close to his chest, and acted privately very differently to his public face. He does not present a reckless gambler hell-bent on risk taking, but if anything, it is a clichéd analogy. Among presidents, Truman was probably the most devoted to poker as a relaxation.[33] Eisenhower was primarily a bridge player: someone who played in partnership, though generally leading the pair. Indeed, in *At Ease*, Eisenhower recounts that in 1920, during his time at Fort Meade, "I decided that I had to quit playing poker. It was not because I didn't enjoy the excitement of the game—I really love to play. But it had become clear that this was not the game to play in the Army. Most of us lived on our salaries. Most losers were bound to be spending not only their own money, but their families'."[34] This points to Eisenhower's innate sense of responsibility. He liked to relax, but not at the expense of those for whom he cared. It also accurately captures his fiscal conservatism. Whether personally or professionally, he had a strong dislike of debt. There is some truth in Thomas's portrayal. However, he rather stretched the analogy too far in search of a winning story that is substantially different from its 2012 publication rivals from Jean Edward Smith and Jim Newton.

While this section is focused on historical appraisals, it must be acknowledged that several works on Eisenhower sit on the cusp of learned journalism rather than wholly academic scholarship. *Eisenhower: The White House Years* by Jim Newton is very much a journalist's portrayal of Eisenhower's two terms in the White House. It is focused on a fast-paced narrative, and fits snugly within "Hidden Hand" revisionism.[35] Newton, like Thomas, built his study around Eisenhower's "Precarious pursuit of peace," and begins his study by focusing on how President Eisenhower managed a May 1, 1958, National Security Council (NSC) meeting.[36] As Newton told it, Eisenhower was assailed by Secretary of Defense McElroy, by his top military advisers and by Secretary of State Dulles to reorient US defense policy from massive retaliation to a more flexible response based on tactical, first strike, nuclear weapons. Yet despite the strength of argument from around the table, Eisenhower was unmoved. He meticulously challenged the utility and plausibility of switching to flexible response, stating it would transform weapons of deterrence—a key aspect of his "New Look" defense policy—into a lightning rod drawing fire onto the United States. He noted too that any change—to either build solely tactical missiles or to build both tactical missiles and a deterrent shield—would vastly increase the defense budget. According to Newton, that would wholly undermine the balance of national security and economic stability and doubtless, create a garrison state. Remaining gracious towards his subordinates, Eisenhower moved the debate on. Nothing said in that NSC meeting made the daily papers and while the meeting was a very significant assertion of the president's power to prevail even over his most senior advisers, there was no hint of a difference of opinion in the public domain.[37] This was how Eisenhower operated: low-key but persuasive; definitely in charge of policy and demonstrably on top of key issues—most especially when they related to economic stability, national security and, most of all, the combination of both.[38] Yet what is frustrating about Newton's work, and indeed all of the modern biographers, is that they are largely re-presenting and re-enforcing old ideas rather than offering a new perspective.

The Eisenhower portrayed in Jeffrey Frank's *Ike and Dick* is rarely gracious and significantly more cunning than statesmanlike. He is also vain, rather selfish, high-handed and aloof. Unfortunately, for scholars anticipating new insight into the presidential character of Eisenhower, Frank's portrayal is also two-dimensional and a rather sketchy caricature. Despite its title, Frank's book is very much a study of Nixon with Eisenhower the set-piece cipher providing the object for Nixon to rail against through the travails of his vice presidency. The Eisenhower who too rarely emerges into any attempt at a more rounded study by the author is modelled on the Schlesinger/Neustadt model: spending rather too much time on the golf course or in thrall to "The Gang" than creating a legacy for the Republican presidency through active support of Nixon. Certainly Nixon's relationship with Eisenhower was remote, and Frank puts the vast majority of blame for that on Eisenhower who appears crueler and even more vindictive than in other portrayals. For the most part, Frank simply does not provide any depth of analysis of Eisenhower's motives, but he does provide occasional vignettes that provide some insight into Eisenhower's modus operandi in the White House.

He noted for instance Eisenhower's insistence that Nixon do his "dirty work" in getting Sherman Adams to resign his post as chief of staff over the Goldfine scandal.[39] Eisenhower was averse to delivering bad news to his subordinates, and always worked to find a way to hand-off the responsibility to others. This was not indecision, but a means to never have to be seen to be the "bad guy." Yet the result was that actions often took longer to happen than if the president had simply taken the initiative in the first place. Equally, if Eisenhower was not entirely convinced of the ability of a subordinate to act with the fortitude, integrity, and duty he expected, he was slow to endorse them—often to the point of frustration of the individual looking for approval. Frank detailed the lengthening list of opportunities from 1952 right through to the 1960 election where Eisenhower could have come out in support of Nixon's presidential credentials, but could never quite bring himself to fully endorse his vice president.

The most recent arrival in the Eisenhower historical biography canon is Paul Johnson's *Eisenhower: A Life.*[40] This brief primer does much to reinforce Eisenhower revisionism, but in such a short narrative, adds little of substance. Space policy is a passing mention, not a subject for analysis.

The historian who has taken Eisenhower revisionism to a new level of insight in recent years is David Nichols. First in *A Matter of Justice* and then in *Eisenhower 1956*,[41] Nichols revisited the detail of two of the most difficult periods in Eisenhower's presidency to focus on how he dealt with crisis.[42] The president's character as leader emerged in far more rounded fashion than the stereotypes of the first wave of Eisenhower scholars, the journalese of some of his modern biographers or even the caricature of Frank. In *Eisenhower 1956,* Nichols restated that Eisenhower maxim: "plans are worthless, but planning is everything."[43] It cuts to the core of the Eisenhower leadership which was essentially, "no surprises." Nichols noted that Eisenhower had done his policy homework on the Middle East long before the tensions over Suez became a crisis. While the actuality of what happened on the ground in Egypt probably sat outside any of his scenario planning, he was confident to act, believing he understood the context both of Nasser's actions and of those of the British, French, and Israelis, and that he understood the consequences of forcing an apparent rift with those countries which, in any other circumstances, were America's staunchest allies. What Nichols conveyed most was that Eisenhower would not be deterred from the strategic vision that he had brought to the White House in 1953. In terms of foreign affairs, this was an unequivocal commitment to the containment of the Soviet Union and Soviet-inspired communism.[44] His action over Suez was entirely rational and in keeping with his commitment to the Cold War "big picture." He was prepared to be cold and prepared to be ruthless in dealing with his supposed allies if it was for the good of the United States and enabled him to keep atomic peace in an era of danger. By using international law to support Nasser's nationalization of the Suez Canal, Nichols suggests "Ike skillfully positioned American power to contain communism, presided over the demise of European colonialism, sought to preserve western access to scarce resources and championed some degree of justice for smaller nations through the United Nations."[45] It is a view that may prove rather too laudatory of Eisenhower's presidential skills in implementing foreign policy, but is a reasonable example of the current state of Eisenhower revisionism.

The complexity of Eisenhower and apparent contradiction between the sharp strategist and occasionally ineffective leader is highlighted in Nichols' study of Eisenhower and civil rights. What emerges here overall is a more cautious leader, one who held a somewhat paternalistic view of managing the country through change, and one who chose to uphold the Constitution and the legal interpretations of it rather than crusade for desegregation for instance. He was publicly ambivalent in his support for the *Brown v Board of Education* decision and slow to take action against Governor Faubus in Little Rock. But his strategic imperative is never in question: his declaration to Earl Warren that he would appoint him to the first vacancy on the Supreme Court even before he took office was "an extraordinary step,"[46] according to Nichols, while his appointments of known integrationists to the Supreme Court appeared at odds with his natural sympathy towards the South, and may indeed have played a part in Nixon's failure to win the south in the 1960 presidential election. Yet even in his Supreme Court appointments there is a strong sense of Eisenhower being powerfully in control. While definitely a political manipulator in his appointments, Eisenhower did not like to be challenged on those appointments, even by his close friends. Following the Warren appointment, Nichols cited a letter written to Eisenhower's long-time Abilene friend, Swede Hazlitt who saw Warren's appointment as overtly political. "It was most emphatically not," Eisenhower replied. "I could not do my duty unless I appointed a man whose philosophy of government was somewhat along the lines of my own."[47] In many ways, Warren was a startlingly left-field appointment as chief justice: a politician with little experience of practicing law. Yet the insight Nichols provided could be applied across many presidential appointments: Eisenhower was most comfortable working with those where he felt there was already a little empathy and common ground.

Nichols explored the sense of duty that underpinned Eisenhower's progress through public life. But quoting Sherman Adams, he also noted that there was significant self-confidence and perhaps implicitly, vanity in Eisenhower's duty. On his decision to run in the 1956 presidential election, Nichols wrote:

> Eisenhower had frequently cloaked his ambition in a soldierly call to serve his country. Sherman Adams, years later, put it more bluntly. "The real reason a President wants to run again is because he doesn't think that anyone else can do as good a job as he's doing." That analysis fit Dwight Eisenhower.[48]

EISENHOWER IN THE POLITICAL SCIENCE EYE

Neustadt was, of course, a political scientist, but one who engaged closely with recent history and had a foot placed firmly in both camps. But there were other influential voices commenting on Eisenhower as president, even while he was in office, from a political science perspective. Written towards the end of Eisenhower's first presidential term, C. Wright Mills's *The Power Elite*[49] offered a broad and systemic sociological insight into the organization of mid-twentieth-century US society. It stated that a single "elite" comprising the hierarchies of state, major corporations, and the military effectively ran America. Mills had been on the faculty of Columbia University in New York when Eisenhower was the institution's president in 1948, and shows some insight into the man he knew as "President" in two senses. Pertinent to this study in particular is Mills's depiction of The Political Directorate.[50] Political decision makers and influencers are discussed in the context of celebrities, the "big rich," admirals, generals, and corporate executives.

Mills saw the political establishment under Eisenhower as being "tightly knit" and "enlarged in scope" with increased power and with a far greater focus in decision making at federal level than ever before.[51] It is notable that Mills regarded Eisenhower as a political outsider: "Occupationally framed by non-political experience."[52] He saw him as closer in style and modus operandi to the military than to the political players and bureaucrats who both served him and sought to advise him, although it was probably not a huge leap of the imagination for Mills to land on such a definition when discussing a president who had spent all but two years of his adult life in the army. Eisenhower's successor, John F Kennedy, by contrast, fitted Mills's definition of a "Party Politician"—with private means and at home with synthetic celebrity.

Mills regarded the executive branch at the center of initiatives and political decision making—taking the three-pronged US political system out of balance (towards the executive) more than ever before. He saw Eisenhower making decisions through "expert counsel and advice"[53] primarily from other political outsiders. He wrote:

> The administration, in fact, is largely an inner circle of political outsiders who have taken over the key executive posts of administrative command; it is composed by members and agents of the corporate rich and the high military in an uneasy alliance with selected professional party politicians, seated primarily in the Congress, whose interests and associations are spread out across a variety of local societies.

To a degree, what Mills wrote was true. Certainly in developing space policy—the concern of this study—Eisenhower's most compelling advice came not from his core political cohort or even his military advisers, but from his scientific adviser, James Killian, and the small group of trusted scientists within the President's Science Advisory Committee (PSAC) that Killian headed. It was a significant sign of his pragmatism: a belief in listening to subject experts rather than those with their own specific personal agenda. Yet his actions in interpreting this advice and making decisions for others to carry forward on his behalf were not taken in isolation from his advisers. Secretary of State John Foster Dulles, and Press Secretary Jim Hagerty in particular were always close to the current debate, while all key policy areas were granted full and free debate within National Security Council sessions—though Eisenhower had usually headed off any potential trouble through discrete one-to-one meetings with key actors well before the open sessions took place. Some years after leaving office, Eisenhower commented on a paper written by his national security adviser, Bobby Cutler, entitled "Use of the NSC Mechanism." He wrote: "members of the NSC became familiar not only with each other, but with the basic factors of problems that might, on some future date, face the President."[54] Essentially, he used the NSC to share and socialize issues that might become problems. The normal format was for Eisenhower to request, via the NSC's executive secretary, James Lay, a plan or report on issues of national security interest. These were generally discussed by all parties in the NSC so that if any of the issues ever became real security problems, there would be plans in place to deal with them but, more importantly, the key NSC actors would be well-versed in the issue and confident to deal with it as the need required.[55] In her essay: "The 'Top of Policy Hill': President Eisenhower and the National Security Council," Anna Kasten Nelson captured the spirit of Eisenhower's engagement with the NSC.[56] "It is clear that the President actively engaged in the meetings and often turned the discussion towards those questions that personally interested him."[57] He clearly found the sounding board useful, but contrary to his early scholarly critics, he did not allow the NSC members to dictate his national security policies. Indeed, she also noted that the NSC and PACGO meetings that followed it was more a means of sharing issues than making crucial national security decisions. "The Eisenhower style was a continuing mix of formal procedure with informal meetings or conversation. The President would meet formally with the members of PACGO (the President's Advisory Committee on Government Organization),[58] but he would precede such a meeting with conversations over breakfast with his brother Milton or with Nelson Rockefeller, both members of the committee."[59]

One of the areas Mills considers is presidential innovation and notes that to be an innovator the president required an inner circle of non-political advisers who mediated between the president, the legislative branch and other outside groups. Clearly space policy came under this theme of innovation—but the book was published a year before the Sputnik mission, and Mills made no specific reference to Eisenhower's actions on space policy.

Perhaps the best known treatise on Eisenhower's decision making in the White House is Princeton political scientist Fred Greenstein's *Hidden-Hand Presidency*.[60] Written just before Kasten Nelson's work, and with access to Eisenhower's presidential papers for the first time, Greenstein set out to revise the prevailing view of Eisenhower as a political amateur.[61] Instead, Greenstein, as has been picked up by Evan Thomas, concluded that Eisenhower was a sophisticated political decision maker who combined the skills of mediation and leadership he had learned in the army with the shrewd demeanor of a seasoned poker player. Indeed, Greenstein suggested he played politics much like he played poker: with a "Hidden Hand" that was much better than his opponents assumed.

Greenstein set out five facets of Eisenhower's political decision making. First, while he was actually a skillful politician, he chose not to let others realize the fact. While he was frequently partisan in his actions, he disguised this by delegating responsibility to others at cabinet, agency, advisory or, on rare occasions, congressional level to implement decisions on his behalf.[62] By deflecting the political element in his decision making, he was able to rise above party politics and both preside and govern as the "President of all the people." Second, Eisenhower was prone to speak in an evasive, convoluted, and often confusing manner. Critics saw this as indicative of a political strategy that enabled Eisenhower to distance himself from unpopular positions on controversial issues—again, it was a means to "fool" his opponents into underestimating him.[63] Third, Eisenhower's man management was built around avoiding personality clashes. He did not make enemies, and even when he disagreed vehemently with a person, he masked his true feelings and got on with the job. This had the additional positive spin-off of helping to maintain his mass popularity.[64] Next, Eisenhower understood how to see issues from his opponents' standpoint and worked on moves and tactics that would win them round to his perspective.[65] Finally, he was a master in the art of delegation. Eisenhower would place important assignments in the hands of subordinates, but never fully loosen the reins of strategic control. If the assignments proved successful, Eisenhower was the first to share the success with those who contributed towards it. But when things went wrong, he tended to disassociate himself from the failure.[66] Greenstein does not tackle space policy directly, but provides a framework for assessing and judging Eisenhower's maneuvering as he set out to distinguish civilian space exploration from missile and surveillance endeavors. Speaking to Greenstein in New York in March 2013, I asked him why he had not written specifically about the Sputnik—NASA period in his study since it appeared to be a prime candidate for a case study illustrating each of the five criteria he had identified in his "Hidden Hand" study. "This was a political science work, not a study of all the issues and aspects of Eisenhower's presidency," he responded.[67]

However, Greenstein's revisionist approach to Eisenhower is incomplete. It created an almost benignly Machiavellian figure, abreast of everything that matters and continually one step ahead of not just political opponents but also his colleagues and advisors.[68] Greenstein provides the political science framework for this research, but this book provides an interpretation of Eisenhower that is rather more human than the model president defined by Greenstein, a model he wanted others to revise and flesh-out.

Writing with David Callahan in 1993, Greenstein did finally tackle Eisenhower's relationship with space policy in "The Reluctant Racer," their essay within Roger Launius and Howard McCurdy's *Spaceflight and the Myth of Presidential Leadership*.[69] They concluded that Eisenhower's reluctance to enter a space race with the Soviets reflected his overall conception of how to fight the Cold War—engaging only from positions of strength. Matching Soviet space efforts might increase US prestige, but would only do so at massive cost and with no material value to either the US economy or national security. The piece contrasted Eisenhower with the seemingly more active Kennedy and follows the traditional line that Sputnik was the catalyst for action and, indeed, that NASA was a creation Eisenhower did not want. But, in allying space policy with Greenstein's view of Eisenhower as a strategic player in presidential decision making, the piece began to cement the idea both that Eisenhower's decision making after Sputnik was important and, probably for the first time, that it was justifiable. However, in taking the traditional path of Eisenhower *reacting* to Sputnik, Greenstein has left a significant gap to be explored.

It is clear from the number of Eisenhower-inspired titles arriving in the 1980s, that the works of Greenstein and to an extent Kasten Nelson were a significant challenge to the orthodoxy. "Eisenhower, A Reputation in Transition," written by David MacIsaac, covered much of the same ground as Schlesinger, but differed significantly in that MacIsaac highlighted two personality traits of Eisenhower that color views of his presidency even today. He noted those traits as: "One was his preference never to be seen in what he did; the other, his lifelong rule to refuse to discuss personalities, to focus all discussion on the issues rather than the people involved." [70] This is a very different interpretation of Eisenhower from the writing of the 1960s. Here is not the benign bumbler, but a much more controlled and disciplined character who would not be drawn into personality politics. Six years later, Robert Burk produced a useful synopsis of the rehabilitation of the Eisenhower reputation in "Eisenhower Revisionism Revisited," yet even by 1988, Greenstein's "Hidden Hand" perspective had become dominant. Others offered cautiously different interpretations: Robert Griffith, for instance portrayed Eisenhower as "the manager of a 'corporate commonwealth.'"[71] This was a position close to William Leuchtenberg's appraisal of Eisenhower as skilled in the arts of bureaucratic management and dedicated to conservative goals.[72]

Greenstein's work has been the subject of irregular revisiting by scholars. On the academic front, *Presidential Studies Quarterly* devoted its Spring 1994 edition to the theme of Eisenhower and Governance.[73] Greenstein, again, dominated the argument, with an updated appraisal of Eisenhower. It was a short article not aiming to offer anything new, but his paper presented a neat analogy: Eisenhower the public head of state and private prime minister.[74] His argument was again that Eisenhower was a leader not merely a figurehead operating to the instructions of John Foster Dulles and Sherman Adams, and much of the article reflected how a growing number of both scholars and the media were now reflecting that interpretation. The Sputnik Fall of 1957 was not a subject of discussion in the issue, though Thomas Gaskin briefly touched on it in a discussion on Eisenhower's relations with Lyndon Johnson on foreign policy. Gaskin's argument focuses on how Johnson used Eisenhower to grow his reputation as a foreign affairs expert. Although Gaskin took a revisionist stance on Eisenhower's response to Sputnik, he placed Johnson at the heart of subsequent policy making in the wake of the satellites. That again is open to challenge.

Perhaps the most interesting observation from the 1994 *Presidential Studies Quarterly* essays was that of Ken Collier who, in his discussion of Eisenhower's relationship with Congress, coined the term "Autopilot Presidency."[75] While Collier's focus is on the "invisible hand" of Eisenhower in terms of congressional relationships—a contrast to Greenstein, and an evocation of Adam Smith, he noted that Eisenhower's method for dealing with Congress was:

> to create a mechanism for maintaining friendly relations with Congress, point it in the right direction, and let it run, taking personal control only during critical moments or during turbulence.[76]

This observation is pertinent to any discussion of Eisenhower's actions following the Sputnik Fall, most particularly in relation to implementing space policy once NASA was operational.

There is another political science angle worthy of comment, and that is how Eisenhower's actions over space fit within some of the key models of presidential effectiveness. Indeed, three connected approaches offer insight into Eisenhower's presidential style without him actually being the subject of discussion. While he used a different case study and a different president to articulate his models, it is impossible not to consider Graham Allison's *Essence of Decision*[77] as offering an illuminating and potentially applicable model for Eisenhower's behavior. Indeed, one suspects that Allison could have written a companion book covering the Sputnik crisis of 1957 to his work on the Cuban Missile Crisis. It would be enlightening to consider his view on Eisenhower's low-key response to what was largely a psychological threat. Tantalizingly, Allison never refers to Sputnik and there is only the briefest mention of the 1960 U-2 incident[78] where Khrushchev and not Eisenhower is the focus of his argument. But it is worth considering Eisenhower's actions within the three models Allison describes.

This study shows that his governmental policy model can, in part, be applied to Eisenhower's actions in devising and anchoring the first United States' space policy—if one is prepared to consider the period 1954–1958 as the timeframe for the policy creation, rather than the commonly-viewed period of Sputnik 1 (October 1957) to the Space and Aeronautics Act (July 1958). The chief descriptors of this model are that: a nation's actions are best understood as the result of politicking and negotiation by its top leaders; that even if they share a goal, leaders differ in how to achieve this because of external factors including personal interests and background; that even the president of the United States must gain consensus with his underlings or risk having his order misunderstood or, in some cases, ignored; that consequently, a leader's entourage will have a large effect on the final decision (i.e., an entourage of "yes men" will create a different outcome from a group of challenging advisors—as Eisenhower had during the period post-Sputnik). The model defines leaders as having different levels of power based on charisma, personality, skills of persuasion, and personal ties to decision makers. It amplifies this by stating that if a leader is certain enough, he will not seek input from his advisers, but rather, approval. Likewise, if a leader has already implicitly decided on a particular course of action, an adviser wishing to have influence must work within the framework of the decision the leader has already made. Next: if a leader fails to reach a consensus with his inner circle (or, at least, the appearance of a consensus), opponents may take advantage of these disagreements. Therefore, effective leaders must create a consensus. And finally, because of the possibilities of miscommunication, misunderstandings, and downright disagreements, different leaders may take actions that the group as a whole would not approve.[79]

Eisenhower has been characterized by Greenstein and others as a consensus builder: but one who built agreement based on his own terms and framework for reference. The thrust of this study is that Eisenhower was focused on a course of action built on learned experience following the H-bomb nuclear tests, the development of the U-2 and latterly the CORONA space satellite program that would have led to a civilian space agency irrespective of Khrushchev's Sputnik power play. How that action manifested itself was influenced by actors outside the main thrust of government—notably Killian, Bissell, and, to a lesser extent, Hagerty and Glennan. Yet these were all hand-picked advisors operating within the presidential mandate. They were a skilled entourage delivering both a framework and detail that Eisenhower was happy to endorse, but the overarching driver was the wish to enhance national security, and that came from the President himself. Equally, one could argue that Allison's Organizational Process Model does not fit Eisenhower's space policy making since he simply would not allow himself or his trusted advisors to be drawn into viewing the launches of the Sputnik satellites as a crisis to which he must respond.

Yet if one looks at Eisenhower, rather than the government as a whole within Allison's models, one could make a case that Eisenhower is a "Rational Actor." Briefly, Allison proposed that the fundamentals of the Rational Actor model were that governments are treated as the primary actor; and that the government examines a set of goals, evaluates them according to their utility, then picks the one with the highest payoff. Substitute Eisenhower for "Government" and one might view his course from 1954–1958 as defining national security as his "goal" in space; seeing enhanced reconnaissance as the means to deliver this and focusing on enabling that to happen to secure the "payoff." This appears to reduce the "white" civilian International Geophysical Year scientific space program and later NASA unmanned and manned programs solely to diversionary status—and, to a certain degree, that is true. But they fulfilled secondary goals of scientific advance and national prestige that probably meant more outside Eisenhower's science advisory circle than within it.

Amy Zegart built on Allison's Organizational model. Her analysis of Eisenhower's failure to effectively reform the Joint Chiefs of Staff (JCS) in *Flawed by Design*[80] is instructive in understanding why the president chose to bypass the armed service route when it came to establishing NASA in 1958. The ease with which covert activities flourished with minimal control is also discussed at length by Zegart, and an argument can be made that Eisenhower entrusted his key reconnaissance projects to the Central Intelligence Agency (CIA) specifically because they would be subject to almost no congressional scrutiny. Zegart's new institutionalist theory focused on the fact that national security agencies, including the CIA, which was her specific focus, were very different from domestic agencies that were characterized by weak interest groups, secrecy, executive domain, and connected bureaucracies. Consequently, they developed differently from other bureaucracies, and were much harder to reform, not least because they remained in the executive domain, largely outside the reach of the other branches of government. Eisenhower failed to reform the JCS effectively. Thus in aiming to achieve his goals for a national security-led space program, he simply chose to side-step the still-powerful armed services chiefs.

The other model to consider is that put forward by Irving Janis in *Groupthink.*[81] Janis had described the key principle of groupthink as:

> The more amiability and esprit de corps there is among the members of a policy-making ingroup the greater the danger that independent critical thinking will be replaced by groupthink, which is likely to result in irrational and dehumanizing actions directed against outgroups.[82]

Eisenhower is not a case-study subject in Janis's work, although he is obliquely referenced in discussions around the Bay of Pigs in relation to the decision to the 1960 U-2 incident.[83] But was he a victim (or perpetrator) of groupthink in relation to Sputnik? The essence of groupthink is that the desire for conformity delivers irrational actions.[84] In taking the position of the first Eisenhower scholars, it would be possible to paint this picture for Eisenhower and his key advisers. When the logical response to Sputnik would have been a powerful military reaction (at least in terms of boosting spending on US military forces) and a battle for prestige through the media, Eisenhower chose a different course. However, the essence of this book is to offer an alternative interpretation of his actions, and will certainly challenge any thought that Eisenhower was a perpetrator of groupthink over Sputnik. However, there is also a case to judge as to whether his advisers operated in groupthink conformity in their actions in creating NASA. Did they deliver a solution that was not fit for purpose? One must argue otherwise. In his book, Janis argued that Kennedy's advisers were sufficiently robust and single-minded to help the president develop effective strategies, particularly in relation to the Cuban Missile Crisis. Johnson's advisers, on the other hand succumbed to groupthink, most especially and disastrously over the escalation of the conflict in Vietnam. Eisenhower would appear to be closer as a subject to Kennedy than to Johnson in his means of decision making over Sputnik.

THE HISTORIOGRAPHY OF EISENHOWER'S SPACE POLICY

It is notable that the traditional interpretation that Eisenhower's actions on outer space were catalyzed directly by Sputnik features in all the most recent works on Eisenhower's presidency. It is also notable that space policy is relegated to minor event status in most recent writings on Eisenhower. Jim Newton, for instance, offers just fewer than three pages out of a 350-page presidential biography to discussing the Sputnik-Mercury narrative.[85] Newton sets up Sputnik 1 as just one of a number of troubles for Eisenhower in 1957. Its launch came as the events in Little Rock, Arkansas, where Eisenhower was forced to use the 101st Airborne Division to ensure the safe passage of black students to the city's Central High School in the face of vehement and aggressive protest from southern segregationists, had finally died down. While the heat had just about gone out of this domestic crisis, the Soviet action over Sputnik added to the discontent of politicians lining up to challenge the presidency as midterms, and then another presidential contest loomed. Newton restated the Sputnik assumption unquestioningly: "The specter of a Soviet eye peering down from the sky sent Americans into an orbit of their own," he wrote. He noted that the launch drove both the latest news from Little Rock and Jimmy Hoffa's election as president of the Teamsters from the nation's front pages, and compared the media's Sputnik frenzy with Eisenhower's sanguine reaction to the launch. Swiftly skating over the intervening months, Newton presents the traditional synthesis of the Sputnik history: "The realization that the Soviet Union could credibly claim that it had surpassed the United States in an arena of strategic consequence roused Eisenhower to action." But rather than investigate what that action was—and why it happened as it did—Newton switches focus to Von Braun and his lobbying of Defense Secretary McElroy.[86] While drawing Von Braun's scheming together with Lyndon Johnson's congressional preening through the Preparedness Subcommittee hearings, Newton paints a scenario where the launch of the Jupiter-C with its satellite payload at the end of January 1958 was a direct response to the two Sputnik launches and was driven by Von Braun and Johnson with Eisenhower somewhat passive in the process. The bill to propose the creation of the National Aeronautics and Space Administration is seen almost as an afterthought, as Newton says nothing about its genesis other than to say Eisenhower proposed it to Congress on April 2, 1958—almost six months on from Sputnik 1's launch. This short-hand speed-telling of the events of late 1957 and the first half of 1958 does little to tease out the nuances that are all important in understanding Eisenhower's role in US space policy. Newton is simply too reductionist in his presentation of the narrative. By way of balance among recent interpretations of the events of 1957 and 1958, Yanek Mieczkowski focuses directly on Eisenhower's response to Sputnik but, as the subtitle to his book *Eisenhower's Sputnik*

Moment makes apparent, does so in a way that regards events as *The Race for Space and World Prestige*. Mieczkowski's *Eisenhower's Sputnik Moment* was the first book for more than two decades to deal directly with Eisenhower's response to Sputnik.[87] It provided a more primary-source driven interpretation that complements Divine's earlier work, but actually does not differ greatly from it.

In *The Sputnik Challenge* published in 1993, more than a decade after Greenstein's revisionism had begun to change the perception of Eisenhower, Divine focused on the impact of the Sputnik launch on Eisenhower's presidency.[88] He argued that Eisenhower completely missed the symbolic significance of Sputnik 1's launch and the Soviets' triumph in the race into space. His core argument was that by tackling the launch in purely rational terms, Eisenhower failed to reassure a fearful public that their fears were groundless. His argument was that Eisenhower's public lack of urgency in responding to the Soviets (while, behind the scenes he was setting the groundwork for a spy satellite program), undermined his leadership and ultimately led to massive rises in military spending and a real missile gap—albeit one in the United States' favor. Divine, who wrote primarily on the twentieth century presidency and foreign policy issues, offers a highly detailed perspective, but one firmly in the "knee-jerk-reaction-to-Sputnik" camp.[89] It was also limited by largely being a secondary source-led monograph. Had Divine chosen to spend more time analyzing primary source material, he may well have come to a different understanding of the president.

The 2013 Mieczkowski book again advanced the claim that Eisenhower was much more a calm poker player in his reaction to events. It is a slightly over-worn analogy. Dividing its coverage into three sections, "Sputnik," "Setbacks," and "Space," *Eisenhower's Sputnik Moment* provided an academic insight not just into immediate events surrounding the "moment," but further into the "space race" through the Nixon-Kennedy presidential race and Kennedy's actions on outer space after his election. In charting Eisenhower's actions from a political perspective, Mieczkowski pushes the case for Eisenhower's activism and sought to positively compare his cautious, moderate and iterative development of a US space program through a comparison of Kennedy's "Race to the Moon."[90] However, he also argued that Eisenhower was politically insensitive in the wake of the Soviet launches, and showed a lack of inspirational leadership in using space as a vehicle to build national and international prestige. Mieczkowski's work is interesting and moves the interpretation of Eisenhower's reaction to Sputnik firmly into revisionist, even post-revisionist, territory. However, on two key counts, its received interpretation is limited. First, it continues to treat Eisenhower as a president reacting to a situation. Second, it ultimately brands Eisenhower as a failure in both politics and perception.[91] The latter point in particular is unconvincing, and an examination of Eisenhower's space policy from the inception of the International Geophysical Year, through to the wake of NASA's creation is required. While Mieczkowski delves deeply into primary source archive material, he does so from a perspective of discussing partisan politics. While he adds to the positive revisions of Eisenhower's second term, he does not challenge the underlying assumption that the president's decision making on space was prompted by the launch of the Soviet satellites.

In terms of assessing Eisenhower using space policy as a framework, until very recently, pro-Kennedy liberal historians held sway vigorously led by former NASA chief historian, Roger Launius. His 2010 paper, "An unintended consequence of the IGY: Eisenhower, Sputnik, the Founding of NASA"[92] adhered strongly to the traditional Sputnik assumption. Launius nailed his colors to the mast quickly with an opening statement that says: "In an irony of the first magnitude, Eisenhower believed that the creation of NASA was a mistake."[93] He then raised several interesting and well-observed points about the manufacture of the Sputnik crisis by interest groups politicking for their own ends. But he displayed an inherent prejudice towards Eisenhower, engaging superficially with Eisenhower revisionism, but largely dismissing it. He wrote:

> [One] can become much less satisfied with the Eisenhower revisionism. Of course, Eisenhower has attained lofty status as some type of grand strategist that seems overdrawn at this point. For example, with American prestige clearly at stake in the Cold War during the 1950s it is puzzling that the chief executive should have been so reluctant to recognize this fact of life. Eisenhower totally mishandled a long list of international intrigues with the Soviet Union, completely misinterpreted the nationalist fervor of former European colonies, displayed alarming incapacity to understand anything happening in Latin America, Africa and Asia, and the list goes on.[94]

Given Launius's rather trenchant position on Eisenhower, it is unsurprising that he sees his actions in 1957–1958 as "unequal to the mentality present from opponents and those they could convince to support them."[95] His observation that NASA was an agency that Eisenhower did not want to create is definitely open to challenge. But one must question Launius's methodology here. He makes sweeping statements concerning Eisenhower's mis-handling of his "long list of international intrigues," yet disdains from engaging with those intrigues or providing any evidence to support his assumptions. The result is a rather lightweight critique limited by boundaries that, given the weight of primary source evidence, can be (and are) breached by a more detailed review of Eisenhower's handling both of space policy and, indeed, those "international intrigues."[96]

Launius has provided a more positive appreciation of Eisenhower's leadership, but only through a review of the work of others in creating a historiography for the origins of the space race. On charting works on what he refers to as the origins of the space age, he commented:

> The figure of Dwight D. Eisenhower has dominated this recent study, and he has emerged as a much more effective leader than thought at the time. Rather than a smiling, do-nothing, golf-playing President, Eisenhower's leadership handling the Soviet Union in space now increasingly appears farsighted and rational.[97]

He cited Divine's *The Sputnik Challenge* as the best analysis of the more active view of Eisenhower and Launius's comment may be read as no more than a reflection of Divine's viewpoint.

Another relatively recent voice in assessing Eisenhower's contribution to space policy is Sean Kalic who chronicled United States' space activity from Truman to Johnson by analyzing presidential actions with regard to the militarization of space.[98] His well-documented though narrow-cast thesis stated that Eisenhower, Kennedy, and Johnson all worked hard to ensure space was not weaponized. All saw military uses for space, primarily for reconnaissance, but also for communication, navigation, and meteorology. With regard to Eisenhower, his specific conclusion was that:

> While the Eisenhower era is often characterized by policies such as massive retaliation and New Look, the period 1953–1961 needs to be remembered as the era in which the United States steadfastly committed itself to the non-aggressive militarization of space, NASA's civilian space program and the global banning of weapons from space. The impact of Eisenhower's actions resulted in a solid commitment to a national space policy that emphasized the use of military and space program working in concert to re-establish the technological superiority of the United States.[99]

Yet Kalic does not explore further whether the Soviets had ever really threatened the United States' technological superiority or whether they had merely given the impression of so doing. Kalic provides just one chapter on Eisenhower in a relatively slim volume and his core focus is heavily on the military aspect of the space program. That said, he points the way to important issues, policies, and sources that need to be tied into the broader debate on Eisenhower's second White House term.

Before Kalic and Mieckowski entered the debate, the interaction of the president, his presiding and operational officers, and the media had been little discussed in the historical analysis of the US-Soviet Space Race. By contrast, the "race" has been probed from almost every other angle as regular anniversaries of the first satellite, first manned flight, first step—and even last step on the Moon have presented opportunities for ever more retellings of the journey from Sputnik to the Moon. Yet what remains startling is how virtually every analysis recycles the core traditional assumption about the impact of Sputnik. The events of the late 1950s are framed as the prelude of a space race that was a Cold War enterprise, and, indeed, a microcosm of the wider struggle. Even a cursory reading of the main literature surrounding early US space policy shows that this is not a debate that easily mirrors the orthodoxy/revisionist/post revisionist convention that one would find when, say, discussing the wider history of the Cold War. The core messages that emerge time and again are focused on the ultimate success of the moon landing both delivering and closing the "New Frontier" as extolled by Kennedy in his presidential nomination acceptance speech at the Democrat Convention in 1960.[100] That space race history has lain largely unchallenged with few exceptions, and those lie very much in the scholarly field rather than making their way into the narrative norms that percolate through wider historical interpretations of this absorbing period in US history. America's early space history is far too often discussed as a stand-alone issue, explained without any contextual reference to the Cold War-dominated foreign policy discourse, or even the US domestic travails of race, economics, and social policy. Additionally, much of the space race literature is in the form of journalistic or memoir-accounts where the retelling of a familiar tale from a new angle is far more prevalent than any insightful analysis.

The first really specific revision to the triumphalist interpretation was Walter McDougall's *The Heavens and the Earth* published in 1985 at a time when the Eisenhower Library was making more and more material from the Eisenhower presidency available to researchers.[101] Published three years after Greenstein's *The Hidden Hand* presidency, his work recast the heroic, triumphal theme that NASA had nurtured by giving a distinct political edge to the rationale of the United States' space endeavors. McDougall focused on the interrelationship of politics and technology and argued that the space agenda emerged as a means to prosecute a war—and especially a Cold War—without that war ever having to get hot. McDougall began to rebalance the political equation, giving Eisenhower due credit for his approach to a peaceful entry into space, and his fiscally conservative approach to managing both the civilian space program under NASA and the defense program both in terms of missiles and reconnaissance satellites. McDougall's approach of historical analysis, drawing on a wide range of primary source materials from NASA, congressional records and the Eisenhower papers, led to his advocacy of the concept of technocracy—the management of society by technical experts.[102] His argument was that the Soviets made it into space first because as a state actor, they were the first to align the political will to technological ability. The United States finally won the race to the moon because they managed the military-industrial complex better and maintained their advantage longer. McDougall argued that Eisenhower was unwilling to embrace the technocratic method, fearing it was out of kilter with his small government philosophy. He was, however, forced to move along the technocratic road by his opponents, prompted by Khrushchev's

Sputnik successes.[103] McDougall still considered Eisenhower to be reactive, and argued that Kennedy was not bound by any of Eisenhower's small government limitations and embraced technocracy fully. Yet Kennedy was no great believer in space and limited his own "New Frontier" by setting the space between the earth and the moon as the most visible symbolic battle ground of the Superpower Cold War. McDougall does not particularly advocate the technocratic approach, but regarded it as the best explanation for understanding what happened between 1957 and 1969. Since writing *The Heavens and the Earth,* McDougall has changed his position, so as to rebalance US presidential achievement in space away from Kennedy and towards Eisenhower. In an exchange of emails in 2011, he noted: "Certainly if he (Nixon) had won (the 1960 Presidential election) and followed the Eisenhower/Glennan NASA plan for incremental progress toward infrastructure the US space program would be decades beyond where we are now, both in manned and unmanned exploration (the latter has suffered because of the former's big budgets)."[104] McDougall is also interesting in that he devotes a chapter to the "Media Riot" that followed Sputnik 1's launch.[105] Yet this is one of the least insightful chapters in his work. The "press" is not deeply analyzed—merely seen as providing a hysterical response leading a hysterical nation which catalyzes Eisenhower into action.

Most serious study of the early American space program in the 1990s and early 2000s was conducted by Roger Launius and Howard McCurdy. Their edited volume *Spaceflight and the Myth of Presidential Leadership*[106] was the first to reassess the leadership of each president through the space program lens. This volume includes Greenstein and Callahan's perception of Eisenhower vis-à-vis Sputnik as discussed earlier.[107] Meanwhile, John Logsdon is arguably the only scholar who spans the small segment of academic literature covering both space and US political decision making. However, his work is almost solely focused on Kennedy and while his access to primary research material is exemplary, there is a sense throughout his work that Kennedy's decision making is not open to challenge. His work spans the period where the orthodoxy dominated, through to the near-present, where revisionists hold the upper hand. Logsdon's 1970 work, *The Decision to Go to the Moon,*[108] was long seen as the definitive study of decision making around the manned space missions. Echoing Allison, it presented Kennedy as a rational actor making the best choice from a range of space-related options. In essence, Logsdon offered the "Moon Pledge" as an instance of the rational choice model of decision making: Kennedy sought to portray the United States as superior to other nations. Entering (and winning) a space race with the Soviets would demonstrate that superiority. Landing a man on the moon and returning him to earth before the decade was out offered the highest return in terms of benefits judged against the likely cost in achieving the aim. Eisenhower's contribution was relegated to very secondary subject matter, with the emphasis being that US space policy only found a purpose through Kennedy's actions. Logsdon returned to the theme in 2010 with *John F Kennedy and the Race to the Moon.*[109] While he added much detail to the argument, he largely stuck by his view that Kennedy's action represented an example of the rational choice model of decision making. Logsdon's research rarely crosses the divide from Kennedy into Eisenhower territory. Again, Logsdon delivers an example of the orthodox narrative of the US victory over adversity in the space race standing out against the mainstream thrust of Eisenhower revisionism. However, in placing Kennedy at the forefront of US space achievement, his work presents an argument open to be expanded. Logsdon does address the issues of space and presidential decision making, but without properly interrogating Eisenhower's activities, principles, drivers, and achievements. As such, the result is incomplete.

A STUDY IN CONFIDENCE

Discussing Eisenhower through the studies of others must be balanced by a review of Eisenhower through his own words. But before that, it is important to acknowledge the core characteristic that is apparent throughout his presidency and in everything he wrote: self-belief. To a large degree, this self-confidence came from his rapid rise through the military in World War II and thus before this book's period of focus. That self-confidence gained the most profound justification through his resounding success in defeating Hitler between 1944 and 1945. Yet he may never have gained that opportunity had it not been for the way in which he caught General George Marshall's attention in the Louisiana Maneuvers of 1941. He was General Walter Krueger's chief of staff when Krueger's Third Army routed General Lear's Second Army in the largest military maneuvers ever held in the United States. Essentially, Marshall saw this as a rehearsal for the real invasion of Europe. According to Smith, Krueger deserved the credit for the victory. "His grasp of the strategic requirements, and the command and control he exercised over Third Army, were nearly flawless."[110] Again, Eisenhower's role was supportive. He was Krueger's chief of staff and thus was the public face of the maneuvers, conducting the daily press briefings. This visibility undoubtedly worked in his favor when officers who had performed well in the exercise were singled out for promotion. In reality, Eisenhower was riding on the tails of Krueger when he was promoted to general. He had performed well in the maneuvers, but the success in the field was very much down to Krueger. This illustrates the view touched on at the very beginning of the chapter of Eisenhower as "lucky." He was lucky in so far as he had a tendency to be in the right place at the right time. He had consistently attracted the attention of his superiors, from Fox Conner onwards, and had performed strongly in each of his postings, enabling him to move, albeit quite slowly, through the pre-war army structure. If anything, he had created his own luck. Being "lucky" is worth nothing if it is not backed by hard work, a comprehensive grasp of the issues in hand, and an ability to deliver at the very least what is required, and often significantly more.

Early in his relationship with his future wife, Mamie, Eisenhower announced his duty would be to the nation first in whatever capacity it chose to use him, and to his family second. It appears that such duty was built on self-control, and that self-control manifested itself in rather few relaxed moments or off-the-cuff comments. A career in army roles demanding discretion, plus a pronounced split between his private and public demeanor, makes Eisenhower a difficult subject to define through his public utterances. So much of what he said or even what he wrote was "for the record," either consciously or unconsciously. Like Churchill, there is often a sense that he is writing for history: to be read by historians and political scholars. With his memoirs, there's a sense of retrofitting action to fit what transpired, while his public speechmaking and even official memoranda of meetings tend to reflect his public character—the measured, smiling, personable character that the first generation of biographers chose to vilify. Thus, in addition to the insight of other historians, and his public statements, there are two ways to capture the character of Eisenhower as presidential leader: first, through the reaction of others to him in the natural interaction of policy making; and second, his less guarded, "off the record" comments. This is occasionally captured in his own thoughts and utterances, but more often through thoughts recorded by others in notes, memoranda and diary entries. Yet getting close to the authentic Eisenhower is not easy. Perhaps wisely recognizing that what goes around comes around, he made it a practice not to speak ill of others, even his adversaries. He had a plaque on his desk with the inscription: "gentle in manner, strong in deed."[111] The "gentle" is reflected in his unwillingness to be openly critical of others, and William Ewald quoted Eisenhower on his philosophy in dealing with those who opposed, or sought to oppose him. The philosophy was simply: "Don't see, don't feel, don't admit and don't answer. Just ignore your attacker and keep smiling."[112] Ewald also noted Eisenhower's self-discipline and highlighted four characteristics he could draw together to achieve his aims: "[He] has an enormous capacity for getting on with people. He could judge them, he could make them like him, and, rarest of all, he could judge himself."[113] What one can surmise from this is that Eisenhower's self-discipline gave him an extraordinary capacity to work with, and draw the best from, people with whom he had no personal affinity. Clearly, that was sometimes a strain and there is evidence from well before he entered the White House that Eisenhower was sometimes impatient, and never had time for the self-important. Indeed, in a rare example of written intemperance, he was highly critical of Douglas MacArthur when serving under the general in the Philippines. He wrote:

> The General is more and more indulging in a habit of damning everybody who disagrees with him over any detail, in extravagant, almost hysterical fashion. I've seen him so this, second hand, in the past, but now he seems to consider that the combined use of his rank, a stream of generalizations that are studded with malapropos, and a refusal to admit the presentation of opposing opinion will, by silencing his subordinates, establish the validity of his contentions.[114]

Robert Ferrell omitted this part of Eisenhower's January 20, 1936, diary entry from the published version of *The Eisenhower Diaries.* Perhaps he felt this outburst of well-written but somewhat angry critique was not in keeping with the image of Eisenhower he wished to present to his readers. However, the diaries do offer occasional insights into Eisenhower's defining characteristics. His White House entries are too few and too considered to enlighten readers about the true Ike, but earlier entries, not least in the period after World War II but before his Republican political life began do show some of the underpinning tenets of the man. For instance, newly-installed as Army Chief of Staff, he writes:

> I'm astounded and appalled at the size and scope of plans the staff sees as necessary to maintain our security position now and in the future. The cost is terrific. We'll be merely tilting at windmills unless we can develop something more in line with financial possibilities.[115]

Here is a man acutely aware of the cost of maintaining a huge standing army in a culture with no tradition of such large-scale full-time soldiery. Even with all the goodwill of a successful European campaign behind him and hero status with politicians and the public alike, his first thought is financial prudence: balancing the needs of the country post-war with the military demands of an army at the height of its powers. It is a view he returns to on May 15, 1947 when taking a more general and discursive view on the strategy necessary for the US economy:

> At home, we have rising prices (Packard had to go up yesterday), labor troubles, housing shortages and tax squabbles. Unless we find a formula which will stabilize the price structure, we are due for most serious times.[116]

It is clear that five years before even running for office, Eisenhower had a strong sense that governmental spending needed to be curbed. While not aligning himself politically at this time, his core economic sentiments were Republican—small government, low federal spending, and support for business, although there is no evidence of a clarion call from Eisenhower for tax cuts. He had also taken a view in April 1946 on the role the army should play to draw the best out of civilian scientific research and development. In a memorandum for directors and chiefs of War Department general and special staff divisions and bureaus and the commanding generals of the major commands he wrote:

> It is our job to take the initiative to promote the development of new resources, if our national security indicates the need. It is our duty to support broad research programs in educational institutions, in industry and in whatever field might be of importance to the Army. . . . The association of military and civilians in educational institutions and industry will level barriers, engender mutual understanding, and lead to the cultivation of friendships invaluable for future co-operation.[117]

Here Eisenhower was showing support for the nation's scientists. Indeed the opening page of the memorandum praises the relationship between the military and its scientific advisers during World War II as a key, but separate, asset for national security. This strong belief in scientific advice and expertise without the clouding of specific agendas sets the tone for Eisenhower's relationship with the science, and specifically the research and development (R&D) community throughout his presidency. Yet the fine line between supporting scientific and technological advances, and becoming in thrall to the unhealthy relationship of the military-industrial complex and to those scientists who placed their own importance above the needs of the nation clearly affected Eisenhower's thinking across the period of restructuring the armed services, managing the post-Sputnik furor and planning his national security legacy. All came together in his final set-piece speech as a national politician—his farewell address. To deliver such an address at all, not least as the election of Kennedy had been a rebuttal to Republican policies, might well be seen as an act of vanity. Yet surely Eisenhower had earned it? He came into the White House a hero, trounced the Democrats twice in presidential elections and kept the nation at peace through eight turbulent Cold War years. At seventy years old, he was bowing out from public office and would not return. This was his last chance to influence public opinion from a national stage. For a figure confident that he knew what was best for the nation, it would have been odd to merely fade from the scene. It is worthy of comment in that it has been so often misinterpreted. As much as this was a warning, it was a recognition that Eisenhower had not been able to rein in the power of the military-industrial complex or to effectively manage (dare I say shackle?) the scientific-technological elite. This might have been achieved under a Nixon presidency, but that was not to be. Thus, in a sense, the farewell address is an apology from the outgoing president who had not achieved all the goals that were clear to him, and could not see those goals being achieved under the incoming administration.

The speech is best remembered for the warning for government to guard against the "unwarranted influence" of the military-industrial complex. But there is more to the speech than this sound bite; more that reflects the fundamental principles of the president. He opened by noting that in three days: "after half a century of service to this country, I will lay down the responsibilities of office."[118] The first reference is to his responsibility: his duty to the nation. This was Eisenhower's abiding sense: service to his country. Next, he reflected on the United States as being: "the strongest, the most influential and most productive nation in the world," but noted that America's "leadership and prestige depend not merely upon our unmatched material progress, riches and military strength, but on how we use our power in the interests of world peace and human betterment." Keeping the peace becomes the major strand of the next section, with a pointed warning that to remain at peace calls for "Not so much the emotional and transitory sacrifices of crisis," and nor should future leaders "feel that some spectacular and costly action could become the miraculous solution to all current difficulties." Here are Eisenhower's principles distilled: peace achieved by long-term strategy, not through short term spectaculars.

The military-industrial complex passage is often taken out of context. Eisenhower was supportive of capitalism and favored small government and the freedom for business to operate. But, he was aware that America had no tradition of a large arms industry or, indeed before World War II, of large armed services. The warning was not to let the combination of "an immense military establishment and a large arms industry . . . endanger our liberties and democratic processes." The section finished with advice: "only an alert and knowledgeable citizenry can compel the meshing of huge industrial and military machinery of defense with our peaceful methods and goals, so that security and liberty may prosper together." This feels as though Eisenhower is putting his strategy of peace through the threat of massive retaliation in the hands of citizenry because he does not trust the incoming administration to follow this strategy. With hindsight, he was prescient.

The last substantial piece in the speech mirrors the military-industrial complex passage and warns both that intellectual curiosity could be lost under the weight of government contracts and, balancing this, that "public policy could itself become the captive of a scientific-technological elite." With his relationship with PSAC in particular, Eisenhower had developed an open, challenging discourse on science that valued and rewarded intellectual curiosity, most especially when it delivered pragmatic solutions. One might draw from the farewell address that Eisenhower feared a situation where policy was hostage to the scientists who had the greatest industrial backing, public pulpit and loudest voices. Referring to that passage, PSAC's Herbert York asked Eisenhower if he had any specific scientists in mind? "And without any hesitation, without thinking, he responded 'Teller and Von Braun.'"[119]

CONCLUSION

Eisenhower's presidential reputation is in a period of relative rehabilitation, yet scholarly opinion of him remains grounded in the "Hidden Hand" interpretation of Greenstein. While the liberal interventionist environment of the 1960s provided the basis for the initial critical assessment of the Eisenhower presidency, there has been no equivalent period of moderate Republicanism during which scholars may have further revised Eisenhower scholarship. Moderate Republicanism began to fall out of fashion, particularly with grass-root activists, almost as Eisenhower left office. The turn to the right, begun under Goldwater was rather slow, but the rise of Reagan and much more recently, the rise of Tea Party politics and even celebrity non-politicians, has left little appetite for scholars to revisit Eisenhower again, despite him being the original celebrity non-politician. That has enabled rather lazy stereotyping to largely go unchallenged. However, there is new interest in Eisenhower, and while biographers such as Newton, Johnson, and Thomas may have little that is new to say, their studies are drawing a new generation of scholars to reconsider one of the most overlooked presidencies of the modern era.

Nichols's approach has been to focus on single issues. First civil rights, then Suez and his forthcoming work on Eisenhower and Joseph McCarthy. This approach has been followed by other Eisenhower scholars, notably Mieczkowski who has explored the Sputnik issue. Yet his exploration has been based on the traditional premise that the post-October 1957 actions were all prompted by the reaction in the United States to the Soviet success in putting a man-made object into orbit first. This orthodoxy does not reflect accurately either the character of Eisenhower or the influence his space policy had on all that followed. Why has this orthodoxy been allowed to survive with little challenge for so long? The first reason is lack of linkage between the space exploration genre and parallel literature on the development especially of reconnaissance satellites, but also military missiles. Eisenhower's space policy is traced back to a slow and underwhelming reaction to the orbiting of Soviet satellites, and especially the initial failure of the Vanguard. Yet this disregards Eisenhower's work with the Technological Capabilities Panel in 1955, to be discussed in the next chapter, which set a clear strategy for both missile and reconnaissance development. One must question why this linkage is not made more effectively. That may be down to the more popular presentation of the United States' greatest success in space—the Apollo program to land men on the moon. Traditionally this narrative has started with Sputnik, but as a failure on behalf of an aging worn out administration. Success comes only when the new administration is in place and is catalyzed by Kennedy's moon pledge. This draws the nation together to deliver a Cold War victory over the Soviets and the ultimate prestige for the United States. It is a triumph for the Democrat administration of the 1960s and a fitting legacy for a slain president. There is a significant element of this narrative that is true. But it is not the whole truth. There is insufficient scholarly research tracing Eisenhower's policy from his first understanding that the United States had the means to deliver its nuclear deterrent via an intercontinental ballistic missile (ICBM), to the implementation of policy that enabled parallel secret reconnaissance and scientific exploration of outer space. Equally, there is insufficient credit in the current literature given either to Eisenhower as a decision maker, or to his legacy in space policy. We might ask: Does the narrative of US space exploration start in the right place? Should scholars be charting US space policy back to decisions made in Eisenhower's first presidential term?

NOTES

1. GK Ericksen, *Women Entrepreneurs Only: 12 Women Entrepreneurs Tell the Stories of Their Success*, (New York, 1999) p. ix. Although Eriksen attributes the quote to Seneca, it may be Seneca quoting Demetrius the Cynic.

2. It would appear that his wartime chief of staff—General Walter Bedell Smith—was the first to attribute the "lucky" tag to Eisenhower, probably in 1945 when the pair visited London and Eisenhower was made an Honorary Freeman of the City of London. In *Eisenhower between the Wars* (New York, 2001) p. 38, Matthew Holland notes Bedell Smith's remark, but does not reference it further.

3. I provide a more detailed evaluation of the Eisenhower and space race historiographies in *Eisenhower's Parallel Track* (Brunel University London, 2014) pp. 19–82, http://bura.brunel.ac.uk/handle/2438/10737.

4. Centers for Disease Control and prevention statistics, 2010, accessed from "Life expectancy at birth, at sixty-five years of age, and at seventy-five years of age, by race and sex: United States, selected years 1900–2007," http://www.cdc.gov/nchs/data/hus/2010/022.pdf, accessed July 26, 2016.

5. Letter from Eisenhower to General Lucius Clay, December 19, 1951, DDE Papers, NATO and the Campaign of 1952, DDE Library.

6. Pulitzer Prize for National Reporting, 1957, James Reston of the *New York Times* for his distinguished national correspondence, including both news dispatches and interpretive reporting, an outstanding example of which was his five-part analysis of the effect of President Eisenhower's illness on the functioning of the executive branch of the federal government. http://www.pulitzer.org/awards/1957. Accessed July 22, 2014.

7. Schlesinger, *Imperial Presidency*, p. 159.

8. E Hughes, *The Ordeal of Power* (New York, 1962).

9. R Neustadt, *Presidential Power and the Modern President* (New York, 1991). This builds on Neustadt's original work: *Presidential Power, The Politics of Leadership* (New York, 1960).

10. Neustadt, *Presidential Power*, p. 139.

11. Neustadt, *Presidential Power*, p. 222.

12. S Hess, conference address at "Ike Reconsidered," Roosevelt House, CUNY, March 7, 2013.

13. Schlesinger, *Imperial Presidency*, p. 158.

14. Ibid., p. 163.

15. A Schlesinger, "The Ike Age Revisited," *Reviews in American History*, Vol 11, No. 1, March 1983, pp. 1–11.

16. Ibid., p. 11.

17. RH Ferrell, editor, *The Eisenhower Diaries* (New York, 1981).
18. JE Smith, *Eisenhower in War and Peace* (New York, 2012).
19. Ibid., p. 177.
20. Following the Louisiana Maneuvers US Army battle preparation exercises of 1941, Eisenhower was transferred to the War Plans Division in Washington, DC under army chief of staff, General George Marshall. His attention to detail in planning led to his promotion to major general in March 1942 (a year previously he had been a Colonel) and his command of the European theater two months later.
21. RM Nixon, "Khrushchev" in *Six Crises* (New York, 1962), p. 363.
22. Smith, *War and Peace,* p. 180.
23. Greenstein notes that while clear expression was natural for Eisenhower, he was not averse to turning to language that was "purposely ambiguous" if it suited his needs—especially to "create smokescreens" to cover actions he had no intention of making public. See *Hidden Hand Presidency*, pp. 66–67.
24. Lucius Clay, Oral History Collection, DDE Library.
25. Niall Palmer, author of *The Twenties in America* (Oxford, 2006), suggests that Harding and Coolidge *both* took Ike's CEO approach, but that the real difference was the "command structure" in the White House, which was more bureaucratic and formalized under Eisenhower with Sherman Adams at the top, but much less hierarchical in the days of Harding and Coolidge.
26. Smith, *War and Peace,* pp. 226–65.
27. Ibid., p. 485.
28. Ibid., p. 470.
29. Eisenhower's correspondence files at the DDE Library are full of letters between the president and members of "The Gang" discussing policy issues. They were also his regular golf partners, at Augusta and other venues, even providing the president and Mamie with a holiday home on the Augusta National Golf course.
30. E Thomas, *Ike's Bluff* (New York, 2012).
31. J Frank, *Ike and Dick* (New York, 2013).
32. Thomas, *Ike's Bluff,* pp. 415–16.
33. Perhaps Truman's most famous evening of poker—on a train ride from Washington, DC, to Fordham, Missouri, in the company of Winston Churchill in 1946 prior to Churchill's "Iron Curtain" speech is recounted by Clark Clifford in CNN's *Cold War* television series—1998–1999. The transcript forms part of the National Security Archive at George Washington University in http://www2.gwu.edu/~nsarchiv/coldwar/interviews/episode-2/clifford1.html. Accessed June 18, 2014.
34. D Eisenhower, *At Ease—Stories I tell to friends* (New York, 1967), pp. 177–78.
35. J Newton, *Eisenhower, The White House Years* (New York, 2012).
36. Ibid., p. 5.
37. Ibid., pp. 2–5.
38. Of course, Newton's assessment of Eisenhower and the NSC is not new, and he draws significantly on the insight put forward in the 1980s by Anna Kasten Nelson, not least in her "Top of the Political Hill" essay discussed in the main text.
39. Frank, *Ike and Dick,* pp. 185–88.
40. P Johnson, *Eisenhower, A Life* (New York, 2015).
41. DA Nichols, *A Matter of Justice* (New York, 2007), and *Eisenhower 1956* (New York, 2011)—2012 paperback version cited in this thesis.
42. At the time of writing, January 2016, Nichols has just submitted the final manuscript of his new study focused on Eisenhower's relationship with Joe McCarthy.
43. Nichols, *Eisenhower 1956,* p. 284.
44. Ibid., p. 277.
45. Ibid., p. 279.
46. Nichols, *A Matter of Justice,* p. 95.
47. Ibid., p. 96.
48. Ibid., p. 120.
49. C Wright Mills, *The Power Elite* (Oxford, 1956).

50. Ibid., pp. 225–41.
51. Ibid., p. 227.
52. Ibid., p. 228.
53. Ibid., p. 229.
54. DDE comments on R Cutler, "Use of the NSC Mechanism," March 1968, Box 1, Gordon Gray Papers, DDE Library.
55. Cutler, "Use of the NSC Mechanism," pp. 2–4
56. A Kasten Nelson, "The 'Top of Policy Hill': President Eisenhower and the National Security Council," in *Diplomatic History*, Volume 7, issue 4, 1983, pp. 307–26.
57. Ibid., p. 311.
58. PACGO—the President's Advisory Committee on Government Organisation—was charged with finding means to implement the decisions made as a result of the papers raised in the NSC meetings.
59. Nelson, "Top of Policy Hill," p. 312.
60. F Greenstein, *The Hidden-hand Presidency: Eisenhower as Leader* (New York, 1982). The revised edition (New York, 1994) has also been used for this book and, where necessary, is referenced accordingly.
61. As portrayed by Neustadt and pre-1980s scholars, most especially in Neustadt's *Presidential Power* (New York, 1960).
62. Greenstein, *Hidden Hand Presidency*, pp. 80–92.
63. Ibid., pp. 66–67.
64. Ibid., pp. 80–92.
65. Greenstein, *Hidden Hand*, pp. 31–35.
66. Ibid., p. 92.
67. F Greenstein in conversation with this writer at the "Ike Reconsidered" Conference, Roosevelt House, CUNY, March 7, 2013.
68. Some scholars have queried whether Eisenhower was a benign figure at all. Stephen Rabe paints a significantly less positive picture of Eisenhower in his book *Eisenhower and Latin America* (Chapel Hill, NC, 1988) where the president is a much more manipulative, than either passive or benign, figure.
69. D Callahan and F Greenstein, "The Reluctant Racer: Eisenhower and US Space Policy" in R Launius and H McCurdy *Spaceflight and the Myth of Presidential Leadership* (Chicago, 1997), pp. 15–50.
70. D MacIsaac, "Eisenhower, a Reputation in Transition," in *Airpower University*, September–October 1982 http://www.airpower.maxwell.af.mil/airchronicles/aureview/1982/sep-oct/macisaac.html. Accessed Sepember 8, 2014.
71. R Griffith, "Eisenhower and the Corporate Commonwealth," *The American Historical Review*, Vol. 87, No. 1, Feb. 1982.
72. W Leuchtenburg, ed., "First Republican Interlude: Dwight D Eisenhower," in *In the shadow of FDR* (Ithaca, NY, 1983).
73. *Presidential Studies Quarterly*, Vol. 4, No. 2 (New York, Spring 1994).
74. F Greenstein, "Eisenhower as Leader," *Presidential Studies Quarterly*, Vol. 4, No. 2, p. 235.
75. K Collier, "Eisenhower and Congress: The Autopilot Presidency," *Presidential Studies Quarterly*, Vol 4, No. 2, pp. 309–25.
76. Ibid., p. 320.
77. GT Allison, *Essence of Decision* (Glenview, IL, 1971) [this was revised in 1999 with Philip Zelikow (New York, 1999) and the 1999 edition is cited in this work].
78. Ibid., p. 242.
79. Ibid., pp. 251–63.
80. AB Zegart, *Flawed By Design* (Stanford, CA, 1999).
81. I Janis, *Groupthink*, Second Edition (Boston, 1982).
82. I Janis, "Groupthink," *Psychology Today* 5 (6): 1971, p. 43–46.
83. I Janis, *Groupthink*, p194.
84. Ibid., p. 242
85. J Newton, *Eisenhower, The White House Years*, pp. 253–56.

86. Ibid., p. 255.
87. Y Mieczkowski, *Eisenhower's Sputnik Moment* (Ithaca, NY, 2013).
88. R Divine, *The Sputnik Challenge* (Oxford, 1993).
89. Robert Divine's work includes presidential biographies of FD Roosevelt and Lyndon Johnson as well as Eisenhower in a wide ranging academic output including works on nuclear testing and the first Gulf War.
90. Mieczkowski, *Sputnik Moment,* pp. 255–79.
91. Ibid., p. 292.
92. RD Launius, "Eisenhower, Sputnik, the Founding of NASA," in *Acta Astronautica* 67, issue 1 (Oxford, 2010), pp. 254–63.
93. Ibid., p. 254.
94. Ibid., p. 259.
95. Ibid., p. 261.
96. DA Nichols's *Suez 1956* certainly provides a strong counter to Launius's interpretation of Eisenhower as a foreign policy actor.
97. R Launius, "The Historical Dimension of Space Exploration: Reflections and Possibilities," *Space Policy* 16 (2000), pp. 23–38.
98. SN Kalic, *US Presidents and the Militarization of Space, 1946–1967* (College Station, TX, 2012).
99. Ibid., p. 59.
100. John F Kennedy acceptance of Democratic nomination for presidential race, July 15, 1960, San Francisco, http://www.jfklibrary.org/Asset-Viewer/u9cdCj1tY0CrCeCxVfZcag.aspx. Accessed April 20, 2012.
101. WA McDougall, *The Heavens and the Earth* (New York, 1985). To a degree, McDougall's work is complemented by Norman Mailer's *Of a Fire on the Moon,* which began as a *Life* magazine series of features giving a first-person journalistic perspective on the launch of Apollo 11, but grew into a single volume existentialist stream-of-consciousness, grudgingly praising the technocratic triumph of the austere and closed world of NASA. It is not about Eisenhower, but captures the essence of the organisation he helped to create—albeit through the filters of the subsequent Kennedy, Johnson and even Nixon presidencies. N Mailer, "A Fire on the Moon," *Life* magazine, (New York 1969) Vol 67, No. 9, pp. 24–41; "The Psychology of Astronauts," *Life* magazine, Vol 67, No. 20, pp. 50–63; "A Dream of the Future's Face," *Life* magazine (New York, 1970) Vol 68, No. 1, pp. 56–74. Collected together as: *Fire on the Moon* (New York), 1970.
102. McDougall, *Heavens and the Earth,* p. 5.
103. Ibid., p. 140.
104. From an email from WA McDougall to Mark Shanahan, March 3, 2011.
105. McDougall, *Heavens and the Earth,* pp. 143–56.
106. *Spaceflight and the Myth of Presidential Leadership,* edited by RD Launius and HE McCurdy (Chicago, 1997).
107. Callahan and Greenstein, "Eisenhower and US Space Policy," in Launius and McCurdy, *Spaceflight and the Myth of Presidential Leadership*, pp. 15–50.
108. JM Logsdon, *The Decision to Go to the Moon,* (Boston, 1970).
109. JM Logsdon, *John F Kennedy and the Race to the Moon,* (New York, 2010).
110. Smith, *War and Peace,* p. 170.
111. The plaque, which sat on Eisenhower's desk in the White House throughout his presidency, now sits on his desk in the DDE museum in Abilene.
112. Ewald, *Crucial* Days, p. 71.
113. Ibid., pp. 21–27.
114. DD Holt and JW Layersaf, eds., *Eisenhower: The Pre-War Diaries and Selected Papers 1905–1941* (Baltimore, 1998), entry for January 20, 1936.
115. DDE, *Eisenhower Diaries,* edited by RH Ferrell (New York, 1981), entry for December 15, 1945, p. 136.
116. DDE *Eisenhower Diaries,* May 15, 1947, p. 141.

117. DDE memo on scientific and technological resources as military assets, April 30, 1946, JRK Book Back-up documents, Box 13, JR Killian Collection, MIT Archives, Cambridge, MA.
118. Box 38, Speech series, DDE's Papers as President, DDE Library—for this reference and the remaining discussion of the farewell address in this chapter.
119. HF York, *Making Weapons, Talking Peace* (New York, 1997), p. 147.

TWO

Missile and Reconnaissance Development under Eisenhower

Too often, especially in the popular narrative, the short-hand story of the United States in space begins with Kennedy's 1961 pledge to send a man to the moon. It is a triumphal story in which the launch of Sputnik back in 1957 is the precursor: the necessary low point from which the United States can emerge only through the vigor of the new president. It is the wrong starting point. The launch of Sputnik 1 doesn't even provide the right starting place. To understand the US space narrative, one has to go back much earlier. Right back indeed to the end of World War II, and the legacy President Truman created for his successor.

When Eisenhower succeeded Truman as president in 1953, he inherited a nation at war. Truman's "Police Action" in Korea had, by January 1953, cost the United States over thirty thousand combat deaths with a further one hundred thousand combatants wounded.[1] By July 27, the Americans, on behalf of the United Nations Command, and the North Koreans, had signed a military armistice, and in terms of dealing with defense issues, much of the rest of 1953 was taken up with repatriating hundreds of thousands of armed forces personnel. The experience of Korea—Eisenhower had indeed followed up his campaign pledge and gone to Korea to see the situation on the ground first-hand—had reinforced in the president a strong belief that "brush fire" wars should be avoided at all costs. His national security focus in his first year in office was on developing a new national defense policy, one that became known as the "New Look." As Eisenhower detailed in *Mandate for Change,* he employed "five basic considerations . . . for designing and employing a security establishment."[2] These were: the assumption that the United States would never start a major war; that modern global warfare would be "cataclysmic beyond belief"; that the relationship between military and economic strength was "intimate and indivisible"; that the armed forces must be modern and not be expected to wage a new war with the weapons of the last; and, finally, that the United States could not be expected to provide all the necessary forces to the free world, but must work with allies.[3] The new policy was a considerable change in direction for a military establishment that had experienced few, if any, limits on military spending since 1941. As Saki Dockrill wrote in *Eisenhower's New Look National Security Policy 1953–1961,* Eisenhower focused not solely on a military response to the global threats the United States faced (primarily, but not totally from Soviet communism), but blended his response with economic, social, and foreign policy concerns.[4] The policy aimed to reshape the potential military response by focusing more on intelligence gathering, covert action and refocusing military capability on nuclear weapons delivered first by Strategic Air Command and, in time, by means of ballistic missiles. Massive retaliation would not be achieved by a large standing conventional force. In short, the primacy of the army was under threat. The policy was embodied in NSC 162/2 which set out the United States' new selective, flexible approach, which relied also on the contribution of NATO allies.[5] There is little doubt that Eisenhower had inherited several significant national security problems: nuclear weapons were expensive; the United States had to keep troops in Europe to allay the fear among the western European allies that they were set to be attacked by the Soviets; and mutual aid was both slow to take off and disproportionately costly to the United States. Dulles, too, with his rather blunt, black-and-white attitude to freedom versus communism, was not quite the diplomatic Svengali that Eisenhower needed in order to get the rest of the free world to play its part. However, as Dockrill noted, Dulles's close relationship with Eisenhower did generate momentum for the New Look. While not entirely supportive of Eisenhower's intention or actions, she reflected that the policy largely worked: Eisenhower created and maintained "a respectable defense posture" while keeping the US economy on track and continuing to rebuild post-war global markets.[6] Eisenhower signed off NSC 162/2 on October 30, 1953. However, a "Basic National Security Policy" that included clauses on: "collecting and analyzing indications of hostile intentions that would give maximum prior warning of possible aggression or subversion in any area of the world;" and "accurately evaluating the capabilities of foreign countries, friendly and neutral as well as enemy, to undertake military, political, economic and subversive courses of action affecting U.S. security," would never be effective if the means to collect the necessary intelligence were not available. Nor was there the means to project US warheads towards their targets in the USSR. As Michael Hogan has discussed in *A Cross of Iron,* the New Look policy was drawn directly from Truman's idea (though never a fulfilled policy) of a capital-intensive containment that relied heavily on air-atomic power.[7] However, during the Korean War, Truman had directed military resources away from missile and satellite research.

The twenty-seven page NSC 162/2 document spent its first five pages outlining the Soviet threat to the United States. It concluded that:

> The authority of the Soviet regime does not appear to have been impaired by the events since Stalin's death, or to be likely to be appreciably weakened during the next few years. . . . The Soviet rulers can be expected to base their policy on the conviction of irreconcilable hostility between the bloc and the non-communist world.[8]

That continued simplistic reading of Kennan's "Long Telegram," and its influence on the Nitze and Acheson-drafted NSC-68 that had been so influential in inspiring Truman's Containment policy appeared unlikely to change, and was reinforced in August 1953 when the Soviets announced they had detonated an H-bomb.[9] However, Eisenhower did briefly consider alternatives including preventative war. First, in a classic, private one-to-one exchange, very much a feature of his preferred style of presidential operation, he asked Dulles by way of a memorandum "To consider whether or not our duty to future generations did not require us to initiate war at the most propitious moment we could designate."[10] Having talked the issue through with his secretary of state and closest political ally, Eisenhower also raised the issue in the NSC meeting of September 24. During the discussion he noted that:

> The United States was confronted with a very terrible threat, and the truth of the matter was that we have devised no way of meeting this threat without imposing ever-greater controls on our economy and on the freedom of our people. We had been trying, in other words, to have our cake and to eat it at the same time. We were engaged, continued the President, not only in saving our money or in defending our persons from attack; we were engaged in the defense of a way of life, and the great danger was that in defending this way of life we would find ourselves resorting to methods that endangered this way of life.[11]

Clearly the options for confronting the Soviet Union were still uppermost in Eisenhower's mind. However, there was no further discussion of preventative war options, and by the NSC 162/2 report, the focus was on other avenues through which the United States would meet that threat. To that end, the report stated:

> The United States must develop and maintain, at the lowest feasible cost, requisite military and non-military strength to deter and, if necessary, to counter Soviet military aggression against the United States or other areas vital to its security.
>
> The risk of Soviet aggression will be minimized by maintaining a strong security posture, with emphasis on adequate offensive retaliatory strength and defensive strength. This must be based on massive atomic capability. . . .

> This strong security posture must also be supported by an effective US intelligence system.[12]

In his memoir of his first term in office, Eisenhower attempted to put the New Look policy into context. He noted that "America could not afford to waste money in any area, including the military, for anything it did *not* need. I knew from experience that there was much duplication among the three services in research and development, in procurement, and even in roles and missions."[13] In fact, even before the New Look policy was signed off in October, Eisenhower had asked Secretary of Defense Wilson to begin a reorganization in the Pentagon to overcome these duplications. In mid-May, the Navy's Admiral Arthur Radford became chairman of the Joint Chiefs of Staff and it was he who actually coined "New Look" for the new defense policy.[14] As well as changes in the Pentagon, the policy demanded reductions in numbers in the army and the navy, not least as these forces reverted to a peace-time standing after Korea. However, the nuclear threat was, for the time being, to be carried to the enemy by the Air Force's Strategic Air Command (SAC), which, over the period December 1953–June 1955, grew by 20,000 personnel. At the same time, the army shrank by a third to a headcount of 1,000,000, while the Navy reduced from 1,000,000 to 870,000.[15] With a policy in place, the grand strategy for defense as defined by Eisenhower, the intent switched from concept to delivery. In this, Eisenhower was helped little by his inheritance from Truman, and rather more by the swift advance in missile development, particularly by the Air Force, and by the interventions of two key enablers for presidential policy, Jim Killian, and Richard "Dick" Bissell, special assistant to the Director of Central Intelligence Allen Dulles.[16]

THE TRUMAN INHERITANCE

The development of the United States' first ICBMs and intermediate-range ballistic missiles (IBRMs) has been the subject of many historical studies. They will not be repeated here. The significant studies of the early ICBMs and IRBMs all argue that little progress had been made by the time Eisenhower succeeded Truman.[17] Following World War II, President Truman had done little to support either the development of rockets/ballistic missiles or of satellites. The focus of the Air Force was on developing the Strategic Air Command (SAC), while the army lost the initiative on rocket development as it allowed the Von Braun team, which had developed the V2 rocket for the Germans, to languish in Fort Bliss, Texas, until 1950, when they moved to the Army Ballistic Missile Agency (ABMA) at the Redstone Arsenal in Huntsville, Alabama. In 1960, Eisenhower stated that Truman's administration allocated just $1 million to long-range missile development in Fiscal Year 1953, sourly noting that this was "less than it was spending to support the price of peanuts."[18] Under Truman, the Air Force was largely obsessed with aerodynamic cruise missiles, and daily was losing ground on the Soviets who had recognized the possibilities of ballistic missile development immediately after the end of World War II when they too had captured a significant number of German rocket scientists and V2 materials.[19] Of course, the rocket engines being developed to propel nuclear missiles to their targets could have more than one purpose. Belatedly in 1952, the government asked Aristid Grosse, a physicist who had worked on the Manhattan Project, to look at the feasibility of building a satellite. His report was not complete until Eisenhower was in office, and it was presented to the assistant secretary for defense, Don Quarles, in the fall of 1953. Grosse concluded that a satellite equipped with a television camera could be a "valuable observation post." But the Air Force advised Quarles that the Grosse report offered "nothing new."[20]

While Truman offered little active support for missile and satellite development, he did not oppose small-scale, small-cost research activities including the research of Project RAND and a small group of Air Force officers—including Bernard Schriever—concerning the use of satellites for military purposes.[21] Project RAND, as it was first titled, was undertaken by the Douglas Aircraft Company's Engineering Division. RAND, a contraction of research and development, had originated in the final months of World War II when members of the War Department, the Office of Scientific Research and Development, and war industries recognized a need for a private organization to connect military planning with research and development decisions. This was precisely the kind of connection that Eisenhower had asked for on April 30, 1946, when he was chief of staff of the US Army.[22] [23]

Published just three days later and sent for circulation among the top echelons of the Pentagon, who were its intended readers beyond its direct Air Force audience, the first RAND report *Preliminary Design of an Experimental World-Circling Spaceship* almost certainly crossed Eisenhower's desk as chief of staff of the army, but there is no evidence as to whether he actually paid it any particular attention before convening the Technology Capabilities Panel.

SATELLITE AND WARHEAD FEASIBILITY

The RAND report presented an engineering analysis of the possibility of creating man-made satellites. It built on existing ideas for rocket technology, proposing designs for both a two- and a four-stage launch vehicle, and concluded that, even in 1946, "modern technology has advanced to a point where it now appears feasible to undertake the design of a satellite vehicle."[24] It also noted: "Such a vehicle will undoubtedly prove to be of great military value," the report stated that the study was centered around a scientific purpose, namely: "obtaining much desired scientific information on cosmic rays, gravitation, geophysics, terrestrial magnetism, astronomy, meteorology and properties of the upper atmosphere."[25] The essential conclusion was that the technology and expertise was there to build a rocket that could put a satellite into earth orbit. What was needed was the will to do so. That was not apparent under Truman. What was also necessary to give the US rocket program the necessary "boost" was a clear purpose. While the RAND report offered a scientific objective, that was never going to provide sufficient reasoning for any president, least of all one so focused on small government spending as Eisenhower, to invest hundreds of millions of dollars in development. The purpose became more clearly defined on March 1, 1954, with the Castle Bravo H-bomb test in the Marshall Islands. This was the first test by the United States of a dry hydrogen device and, rather than the four to eight megaton blast initially envisaged, the device produced a yield equivalent to fifteen megatons of TNT. The Castle Bravo test was designed to evaluate a first generation air-droppable device. The unexpected consequence of the test was that the United States could package a smaller but far more powerful nuclear device than originally anticipated. Indeed, the Mark-21 nuclear device, twelve foot six inches long and weighing fifteen thousand lbs., was operational by the end of 1955. For the missile engineers working on both army and air force projects, it made the possibility of mounting a nuclear device on an ICBM feasible.

Eisenhower had wanted to keep the results of the Castle tests secret. However, the fact that a Japanese fishing boat had been "showered with radiation"[26] prompted a protest from the Japanese government. Eisenhower then dealt with the issues raised by the Castle tests in two press conferences, on March 24 on his own, and then a week later with Lewis Strauss, chairman of the Atomic Energy Commission (AEC). At the first news conference, George E. Herman from CBS Radio asked: "Some anti-American newspapers in Japan and other countries in the Far East, have been seizing upon these cases of radioactive poisoning to make some very strong anti-American propaganda. I wonder if you would care to give us some statement of policy of the Government of its responsibility towards the rest of the world in these tests?" Eisenhower responded: "It is quite clear that this time something must have happened that we have never experienced before, and must have surprised and astonished the scientists. Very properly, the United States has to take precautions that never occurred to them before. Now, in the meantime, I know nothing about the details of this case. It is one of the things that Admiral Strauss is looking up, but it has been reported to me that the reports were far more serious than the actual results justified."[27] Both the question and the response were quite low-key and respectful. A week later, through Admiral Strauss's response, rather than any answer from the president, the story, and the public impact of Castle Bravo, became significantly bigger.

Eisenhower enjoyed news conferences as they enabled him to talk directly to members of the press and thus keep some control over the tone and content of messages emanating from the White House. However, on March 31, he was unable to control Strauss's rather more vivid way of presenting information. Asked by Richard Wilson of Cowles Publications to describe the kind of damage and geographical scale of destruction the Castle Bravo blast inflicted, Strauss responded: "Well, the nature of an H-bomb, Mr. Wilson, is that, in effect, it can be made to be as large as you wish, as large as the military requirement demands, that is to say, an H-bomb can be made as-large enough to take out a city. (A chorus of 'What?') To take out a city, to destroy a city." An unnamed journalist asked: "How big a city?" Strauss responded: "Any city." another question came from the room: "Any city, New York?" To which Strauss answered: "The metropolitan area, yes."[28]

Hagerty's diary recorded that the president was not best pleased with Strauss's answers. He wrote:

> On way down in elevator President said: "Lewis, I wouldn't have answered that one in that way. I would have said: 'Wait for the movie.' But other than that, I thought you handled it very well." Strauss considerably upset. Thought President was mad. Actually, President has habit of reviewing actions with people and expressing what he would have done in similar situation. Just like post-mortem on a bridge hand.[29]

It is interesting to consider why Strauss was upset. He was a significant political player while at the AEC but not one of those scientist-bureaucrats who became a close adviser to the president. Strauss wanted to be seen as a newsworthy figure himself, not part of the supporting cast of experts around the president. As such, he was never a close confidant of Eisenhower, never a true "Helping Hand." However, he played an important role in communicating just how powerful Castle Bravo had been. The message was not lost on the American public—not least as next day's newspapers, including the *Washington Post* reproduced the text of Strauss's H-bomb test statement in full.[30]

If Eisenhower required any precedent to act to draw together the strands of missile development and intelligence gathering, he found that precedent in his inheritance from Truman. In the closing days of Truman's presidency, just a week before Eisenhower's inauguration, Truman had signed an NSC paper noting that the United States' ability to defend itself from atomic attack was extremely limited.[31] The traditional American defense based on geography no longer held weight, and the new president's inheritance was weak development of new missile technology, and a perceived growing threat from the USSR based, to a large extent, on little credible intelligence detailing what was actually happening in bomber and rocket development within the borders of the Soviet empire.

THE TCP

Jim Killian was the fulcrum of what became the Technological Capabilities Panel (TCP). In his memoir, he detailed an intense eleven-month period from first conversations with the president through to the presentation of the TCP's report *Meeting the Threat of Surprise Attack.*[32] This brought Killian into close contact with Eisenhower for the first time (although they had met when both had been college presidents), and also created the blueprint for organizing NASA four years later.[33] It succeeded because its operation and outputs chimed directly with Eisenhower's style and preferences. The TCP provided a cool, impersonal analysis of the United States' readiness to deal with a surprise attack from the Soviets. It operated on almost military lines. It delivered its report when it said it would, and that report was not a litany of issues, but a tightly-worded study leading to actionable recommendations that would enhance offensive attack capabilities, homeland defense, and perhaps most importantly, provide, for the first time, an effective intelligence gathering platform that would bring accuracy to US estimates of the Soviet threat, exemplified by its Bison bomber fleet and its initial ICBM deployment. Donald Welzenbach captured the strength of Killian well, describing how he had gained Eisenhower's respect as a "presiding officer who could draw people together and the constructive solutions to problems."[34]

Eisenhower's call for a study it seems was sparked by a Pentagon report in 1953 that the Soviets had developed a new strategic bomber, the aforementioned Bison, capable of attacking US cities with nuclear weapons.[35] He first met with the Office of Defense Mobilization (ODM) Science Advisory Committee, of which Killian was a member, on March 27, 1954. The ODM was the agency created in 1950 to oversee defense procurement, production, wage and price controls. Killian was an obvious member, having been part of the wartime Office of Scientific Research and Development.[36] By 1953, this had become the ODM's Science Advisory Committee under the chairmanship of Lee DuBridge. Eisenhower discussed the danger of surprise attacks on the United States, expressing concern that "modern weapons had made it easier for a hostile nation with a closed society to plan an attack in secrecy and thus gain an advantage denied to the nation with an open society."[37] He challenged the committee to find ways of reducing the probability of surprise attacks. The response was for the ODM to convene a task force, endorsed by Eisenhower, to look at three areas of national security: continental defense, strike power, and intelligence. There were to be supporting studies in communications and technical manpower.

The attraction of the task force panel for Eisenhower was that it was to be made up of Washington insiders, but not overtly "political" insiders. All the investigators were scientists and engineers, and the steering committee, chaired by Killian, was a blend of scientists and entrepreneurs who had built up technology-focused businesses. They brought to the issues independence of thought and no agendas shaped either by Congress, the armed forces or the growing lobby of military-industrialists. Donald Quarles, assistant secretary of defense, also saw the benefit of an independent study, but found that, despite Eisenhower's instruction, he still had to sell the idea to Secretary Wilson before the panel members could get access to the people and information they needed. He made his case to Wilson stating:

> I have reviewed the proposed communication [to Eisenhower recommending a full study] and believe that it is well thought out by as competent a scientific group as could ever be assembled, who earnestly desire to be helpful in a matter of such great national importance. . . . The value of such a study appears to lie in the fresh, imaginative approach to the facts, and in the spotlighting of the important inferences to be drawn from them.[38]

The early criticism of Eisenhower by Schlesinger and Neustadt was that he was a political naïf; the outsider who could not manage the political aspect of the Executive agenda effectively. But that is a reductionist reading of what it is to be a political player. Eisenhower was a Washington insider through-and-through who had worked closely with Roosevelt through the war and had been closely associated with Truman through his roles as chief of staff of the army and supreme commander of NATO under Truman's Presidency. Earlier, in 1927, he had worked for Pershing in Washington at the American Battle Monuments Commission. From 1929–1933 he had been the assistant secretary of War's Executive Officer. He had followed this as chief military aide to General Douglas MacArthur when MacArthur was chief of staff of the army. Having followed MacArthur to the Philippines, he returned to the United States in 1939, again, briefly to Washington. After a number of staff roles in operational units, he finally joined the general staff in Washington after the Pearl Harbor attack, spending seven months scenario planning before being reassigned to Europe. Eisenhower had not entered the White House by a traditional congressional or gubernatorial route, but he was far more *au fait* with how Washington worked than many other presidents before or after his time in office. And, of course, while working for Marshall in Washington, he had been fully aware of the work of the government's Office of Scientific Research and Development. One should also acknowledge his ability to work with congressional Leaders such as Sam Rayburn in the House of Representatives and Lyndon Johnson and Richard Russell in the Senate. This ability was a pragmatic "strength" of Eisenhower that is generally not appreciated by neo-conservative commentators on the presidency. For six of his eight years in office, Eisenhower worked with a politically-opposed Congress. Yet he achieved most of his strategic aims, not least reshaping the armed forces through a series of budget cuts across his presidency.

There is a certain tongue-in-cheek humor in the way William Burrows presents Killian's actions following the March 27 initial meeting:

> Eisenhower mandated his science brain trust to come up with solutions to the surprise-attack problem, and they, in turn, responded in the time-honored way of academicians the world over. They decided that Killian, then President of MIT, should form a subcommittee to study the feasibility of conducting a full-blown study. The subcommittee reported less than three weeks later that it believed a major study was indeed justified. Killian duly received a letter from Eisenhower on July 26 urging him to head the study. He accepted and had his forces deployed by summer's end.[39]

However, once given Eisenhower's mandate, there was nothing humorous about the TCP's operation. The core of the panel was a steering group headed by Killian and three project teams of investigators, amounting to around forty people in total, and a military advisory committee. They represented the very best of the United States' military, scientific, and industrial communities. Over a period of around twenty weeks, the three investigating teams convened over three hundred meetings with every major US military defense and intelligence unit.[40] With Eisenhower's direct endorsement, they had access to all of the United States' intelligence and defense programs and brought to their analysis an unpartisan, slightly aloof, educated and informed style of questioning. In essence, this was a management consultancy audit of US national defense and intelligence conducted by the best scientific minds in the country. As a consequence, the final report, pulled together by Killian's steering committee in a ten-day period from February 4–14, 1955, had an air of cold, objective authority.[41]

CONTEXT FOR SPUTNIK

As a means of understanding Eisenhower's muted reaction to Sputnik two and a half years later, two sections of the report are crucial. They are those focused on missile development and on intelligence. It is notable that "Project Three," investigating the United States' intelligence capabilities was led by Polaroid President Edwin (Din) Land, who was to lead the camera development on the U-2 and CORONA satellite, and Ed Purcell, who was later to be a crucial actor in defining the scope and operational role of NASA. Both were close to Killian but also fitted the role of useful scientist for Eisenhower. Unlike Strauss, they put scientific service to their country before personal recognition, although that may have been easier given that one headed one of America's fastest growing corporations, while the other was already a Nobel laureate.

Several recommendations of the TCP stand out. First, in relation to missiles, "General Recommendation 2" stated that the National Security Council should: "formally recognize the present Air Force program for the development of an intercontinental ballistic missile as a nationally supported effort of highest priority." Within this, there were two relevant specific recommendations:

1. The development of an intercontinental ballistic missile (with about 5500 nautical mile range and megaton warhead) [should] continue to receive the very substantial support necessary to complete it at the earliest possible date.
2. There be developed a ballistic missile (with about 1500 nautical mile range and megaton warhead) for strategic bombardment; both land-basing and ship-basing should be considered.

Then, in relation to intelligence, section C offers the rather coy statement:

1. We *must* find ways to increase the number of hard facts upon which our intelligence estimates are based, to provide better strategic warning, to minimize surprise in the kind of attack, and to reduce the danger of gross overestimation or gross underestimation of the threat. To this end, we recommend adoption of a vigorous program for the extensive use, in many intelligence procedures, of the most advanced knowledge in science and technology.[42]

That final statement on intelligence said very little, and this was entirely deliberate. The opening of the report had noted: "We obtain little significant information from classical covert operations inside Russia. . . . We cannot hope to circumvent [Soviet security measures] in an easy way. But we can use the ultimate in science and technology to improve our intelligence take."[43]

Eisenhower had met with Killian and Land on several occasions throughout the TCP's data gathering phase and had instructed them to provide a separate "eyes only" report on their intelligence recommendations that, for the sake of secrecy, would not be shared with the whole NSC. It was these reports that provided the basis for the development work on the U-2 and its reconnaissance capability. However, the TCP report in February 1955 did not provide an immediate "green light" for a crash program of missile development nor for work on the new spy plane. Eisenhower remained both innately cautious and conscious of process and policy. Therefore, his immediate action was to issue NSC Action

1355 which asked for comments on the TCP recommendations from all the departments and agencies involved in the NSC by June 6, 1955. These, in turn, were collated as NSC 5522 which was the subject of discussion at the 257th meeting of the National Security Council on August 4, 1955. At this, the president broadly gave the go-ahead for speeding up ICBM and IRBM development, while just one line in the minutes of the meeting record support for the Killian approach to intelligence gathering. Of interest too for the development of future space policy was an action which stated: "C–8: Initiation of program for small earth satellite (being implemented under NSC 5520)."[44]

Zuoyue Wang, who has provided a comprehensive study of the President's Science Advisory team throughout the Cold War, noted that the TCP owed its success to the caliber of scientist prepared to join the panel—especially as it was recruiting at the same time as the investigation into Robert Oppenheimer was underway. He suggests that moderate American scientists saw the panel as an opportunity for them to contribute to national security, maintain their influence and, if possible, to ease the arms race.[45] Now, through a formal review system, recognized and implemented throughout Executive circles in Washington for many years, Eisenhower had found a way to involve catalysts for change, who could cut through the stasis built on interservice rivalry and add impetus to a slow-grinding and dysfunctional military and intelligence planning process.

STRATEGY INTO ACTION

Following the TCP, the development of missiles was finally given the highest priority in September 1955, while work also stepped up rapidly on the development of long-range reconnaissance. Eisenhower adopted differing roles in missile development and in intelligence gathering through reconnaissance. With the U-2, follow-up A-12 (SR-71), and CORONA programs, Eisenhower used the CIA to deliver his needs. The result was a relatively rapid solution that met the president's national security requirements. For missiles, he largely left the army, air force and navy to compete with each other for resources, ideas, and the expertise to turn those ideas into functioning missiles. The result was duplication, delay, and solutions that, by the time of Sputnik, had done nothing but raise the political temperature in Washington.

The prevailing view among scholars is that Eisenhower placed the U-2 and later CORONA projects with the CIA as a means to avoid the slow-moving bureaucracy and infighting of the armed services.[46] It has also been suggested, notably by William Burrows, that if a secret reconnaissance plane was shot down over the USSR, the political blow would be less severe if it was flown by a civilian rather than a military pilot.[47]

However, there were two other prime reasons why Eisenhower wanted the projects to be under the CIA rather than the Air Force: secrecy and control. The fact that the TCP recommended giving highest priority to ICBM and IRBM development could not be hidden when the president chose to act on the panel's recommendation. Elevating the development to the highest priority status demanded additional funding and that had to be approved by Congress. Congressional oversight was largely conducted in the public eye and thus, while the details of the missile development remained secret, the fact that development was happening was very public. This was both politically convenient domestically, and a powerful warning to the Kremlin. The development of a high level reconnaissance program to source intelligence could never be so public. Eisenhower had to find a route for development that maintained strict secrecy, but also that kept the program away from any potential congressional meddling that might either compromise its effectiveness or alert the Soviets to its existence. The "Aquatone" project, as it became known, was hardly the most secret of secrets. Though CIA-led, it relied on support from the Air Force's Research and Development Division. What made it different from other aircraft research projects was that it was achieved beyond congressional scrutiny. While the CIA had to present its budget to Congress each year, significant portions of that budget were beyond the reach of congressional oversight, being deemed for national security needs and therefore under the direct control of the executive, in this case through Allen Dulles as head of the CIA.

Another reason why Eisenhower chose not to take on the armed services at this time in terms of missile development was that he still had an election to fight. As a first term president, every action he took was scrutinized in terms of the upcoming 1956 election. Taking on the Pentagon meant also taking on an industry dependent on contracts to support a garrison economy and, in effect, risking the vote of every defense worker or worker in a defense-related job.[48] While political lobbying was not quite the black art that it is today, the first duty of a member of Congress was to his or her local constituents. And if that constituency saw its economy fueled by a successful defense plant or research establishment, said congressman would work hard to defend that business and its workforce, prompted by well-funded interest groups lobbying on behalf of the defense industry. If he was serious about reelection, Eisenhower would certainly not plan to stir up hostility among his own political supporters. Meanwhile, taking on the defense industry also meant taking on the media that depended on the defense industry for its income through advertising. And whether it was *Collier's* magazine or *Aviation Week*, this was a section of the media that was widely read, widely respected, and politically influential.[49]

The lobbyists, both from inside the Air Force's missile research and development operation and their defense industry partners; and through the congressmen they had already won over, had worked hard following the TCP to earn a face-to-face meeting with Eisenhower. Without this, it is unlikely that Eisenhower would have signed the presidential directive NSC 1433 on September 13, 1955, proclaiming ICBM development "a program of the highest priority."[50] However, the protagonists, notably the Air Force's assistant secretary for research and development, Trevor Gardner; John Von Neumann, the AEC commissioner who had previously chaired the Air Force's strategic Missile Evaluation Committee; and Bernard Schriever, the Air Force general in charge of missile development had used the Killian committee, and congressional pressure to create an opportunity to address Eisenhower directly. Air Force officer Vincent T. Ford, who served on the staff of the Department of the Air Force, Office of the Special Assistant for Research and Development, had been a seconded member of the Killian committee staff, and had remained working for the TCP once the report had been submitted. From his position within the committee reporting to David Beckler, the committee secretary, Ford was able to raise the issue of ICBM development with assistant secretary of state, Robert Bowie, chair of the State Department's policy planning council, and state's representative on the NSC's policy planning council.[51] It was thus, by a rather roundabout means that ICBM development, recommended for priority in the TCP report but still a backwater within an SAC-obsessed Air Force, reached the NSC in July 1955.

The Gardner, Von Neumann, and Schriever presentation offers a perfect study of how to go about winning Eisenhower's support. The presentation, as described in the eyewitness Ford's unpublished manuscript: *Twenty-four Minutes to Check-Mate*[52] both charmed and flattered Eisenhower and presented a solution that would deliver an ICBM before the Soviets did, and within the strict financial constraints decreed by the president. Gardner set the context:

> Mr. President, gentlemen, the Air Force is privileged to have this opportunity to tell you of the work going on at Inglewood, California in the crash development of an intercontinental ballistic missile. The development of such a device as a vital part of our strategic force structure was not considered technically or economically feasible until Dr. Von Neumann and his panel of scientific experts, using the data derived from the CASTLE tests in the Pacific, concluded that it was now possible to develop a nuclear warhead of high yield and low weight, and that delivery of such a warhead by intercontinental ballistic missile is now strategically and economically feasible. This conclusion by Dr. Von Neumann and his colleagues is known as the "thermonuclear breakthrough."[53]

Gardner opened with due deference but quickly cut to the chase, building on what Eisenhower already knew from the Castle-Bravo test. But he approached the topic with a rare tactical intelligence, not presenting his finding as an Air Force achievement, but as the result of independent research by a group of independent scientists. The TCP had shown this to be Eisenhower's proven preferred *modus operandus*. Presenting the Von Neumann committee as an echo of the TCP was an effective opening. Von Neumann added technical gravitas and reiterated one of Gardner's points: that the United States needed to invest in missile technology immediately if it was to beat the Soviets. Gardner had referred to a nuclear armed ICBM as being "of overriding importance to the security and survival of the US" and Von Neumann re-stressed the importance of having this technology before the Soviets.[54] Eisenhower was no believer in an arms race, but the essence of the New Look was to have just enough technology-based weaponry to hold a national security advantage. Therefore, this was another telling point. But Schriever, like Eisenhower a West Point graduate, cemented the Air Force's case. His speech was tailored to Eisenhower's needs. It implied that the Air Force was in close alignment behind the ICBM program (it was not!) with the inference that there would be a clear path for successful development. It presented a very business-like management structure drawing on close collaboration between the Air Force and technical experts—the kind of alliance Eisenhower had envisaged in his 1946 chief of staff memorandum. It advocated modern development methods to deliver the right solution on time. Most of all, it emphasized cost-consciousness, not the wastefulness and duplication of cost and effort the president was used to from the armed forces. As such, it was a tour de force, and was an influential factor in Eisenhower's decision not to appoint a missile czar in 1955.

BISSELL, THE CIA, AND RECONNAISSANCE

Those elements of Eisenhower's strategy that were most successfully implemented tended to be the ones led by people who fitted the mold of an Eisenhower-esque operational officer. Dick Bissell, in 1954 the special assistant to the director of Central Intelligence, met that requirement perfectly. As part of "the Georgetown Set" of Washington insiders, Bissell, an economist by training, was already well-known in executive circles. According to his CIA profile, he earned a strong reputation as an effective planner and project manager during World War II at the shipping adjustment board in Washington. His "economic background, childhood fascination with memorizing train timetables, and superb organizational skills earned him a reputation as '*the* American merchant shipping planner.'"[55] After the war, he taught at MIT, which brought him into contact with Killian who recruited him from Yale, and in 1948, he returned to Government service to help draft and implement the Marshall Plan. First, Bissell served as executive secretary to Averell Harriman on the President's Committee on Foreign Aid, and then as assistant administrator to

Paul Hoffman, the administrator of the Economic Cooperation Administration, which was the executive agency charged with delivering the Marshall Plan. However, there is no particular evidence to suggest that Bissell and Marshall engaged at anything more than a professional level and, indeed, much of Bissell's work for the ECA took place after Marshall's retirement from the State Department. Bissell was, however, deputy administrator and ultimately administrator of the ECA while Eisenhower was in Paris setting up NATO. Each was aware of the activities of the other. Indeed, in an oral history interview, Bissell noted that he had met Eisenhower "before the [1952 presidential election] campaign."[56]

In relation to his work on the U-2, its successor the A-12 and the CIA elements of the CORONA reconnaissance satellite, Killian recorded that: "Bissell proved to be a brilliant project engineer."[57] That was most certainly unexpected since he had spent his entire career as an economics expert. But his capacity to learn and apply new skills made him an invaluable support in turning Eisenhower's agreed strategy into practice. Indeed, this is a point of comparison and similarity with Eisenhower. Both he and Bissell were autodidacts, able to assess and build understanding of new fields with considerable speed. As with Killian, Bissell met regularly with Eisenhower and developed a strong admiration for the president. "Eisenhower was a man for whom I shared the national admiration. I came to know him a bit more (I'd known him before the campaign) in the Presidency and saw him on a good many occasions. I liked and admired a number of men in the Eisenhower administration, especially during its later years."[58] While it is not to say that the admiration was not reciprocated, Eisenhower's abiding obsession with secrecy around matters of national security means that Bissell, this key "Helping Hand," does not appear in his diaries, nor in his memoir of his first term in office, *Mandate for Change* and the only official records of his meetings with the president across the four years from the first work on the U-2 to the sign-off decision on the CORONA satellite are General Goodpaster's memoranda for the record.[59] Yet in the U-2 to CORONA period, Bissell was absolutely vital to US secret reconnaissance efforts. At least one intervention from Eisenhower on his behalf ensured that the CIA man was able to turn the president's strategic intent into action that gave the United States a significant intelligence advantage over the Soviets and ensured Eisenhower knew the full picture of Soviet missile and bomber capability in the wake of the Sputnik launches.

Much of the key to Bissell's successful service to the president came from continuity. In leading the intelligence study as part of the Killian Surprise Attack Panel, Polaroid's Din Land was supported by a team of six, including Jim Baker, the foremost designer of aerial lenses, Ed Purcell, who would later serve as a key PSAC member, and John Tukey who would also become a long-serving PSAC member. Land and Baker in particular, remained active in the reconnaissance developments post-

TCP, with Baker designing the cameras and lenses for the U-2 and later A-12. The rest of the prime team involved in the design of the U-2 and its testing was also far smaller than would have been acceptable under any air force project, ensuring speed, efficiency and the opportunity for radical design throughout the process. This was all enabled by what might be seen as an unusual demand from Eisenhower at the outset that the development be "handled in an unconventional way so that it would not become entangled in the bureaucracy of the Defense Department or troubled by rivalries among the services."[60] The demand might be unusual as Eisenhower was a product of the armed services but wanted the development kept out of their reach. Yet this clearly reflects two aspects of his thinking. First, his demand for secrecy in all matters of national security, and second, his pragmatism when it came to aiding projects of vital importance to national security. There is no mention of Eisenhower's U-2 "demand" in any official White House or NSC papers, and the only corroboration comes in Bissell's memoir, which repeats Killian's assertion, as does Burrows in *Deep Black*.[61]

The most important issues that the U-2 team faced were funding for the most expensive elements—in this case a brand-new airframe; and a secure location where all testing could take place beyond the prying eyes not simply of the public, but of a potentially jealous air force as well. As ever, Eisenhower was prepared to get involved where his intervention was necessary, but his preferred way of working was to have subordinates approach him with a solution rather than a problem. The airframe funding issue was raised by Trevor Gardner, the Air Force assistant secretary for research and development as early as November 1954. The air force was not exactly excluded from the U-2 project, but was confined to a supporting role, not leading on any aspect of the project. In that meeting discussion turned to who would pay for a new airframe, the cost of which would run to tens of millions of dollars. In his memoir, the only written recollection of the event, Bissell recalled:

> I was sitting near the middle of the table. As I turned to my right, everyone was looking in my direction. As I turned to the left, everyone on my left was looking in my direction as well. I got the point pretty quickly and said I would recommend to Dulles that funding for the project be provided from the CIA's contingency reserve, a reserve appropriated and voted on by Congress. The rules for its use were that withdrawals had to be authorized by the director of the budget and approved by the President on the recommendation of the Director of Central Intelligence.[62]

It was a neat solution and an example of hiding in plain sight that would become a feature of Eisenhower's emerging scientific reconnaissance program. While directly mandated by the president, it had tacit congressional support and a cut-out via the CIA that allowed the president an element of deniability (a feature of Greenstein's "Hidden Hand" assessment) should anything go wrong.

Eisenhower also intervened directly in securing a development base for the radical new spy plane. Having scouted out possible secure sites, Bissell recalled:

> When I returned to Washington, I recommended to Eisenhower that he add a piece of adjacent land, including Groom Lake [the dry bed lake airstrip vital for U-2 test flying] to the Nevada test site of the Atomic Energy Commission. The commission's work was already highly classified, and enlarging the site area would be the easiest way to prohibit over-flights of the new U-2 base without arousing attention from outsiders. Eisenhower approved the proposal.[63]

Aside from giving a clear instruction to implement the recommendations of the intelligence work stream of the TCP, an instruction to use unconventional development means to maintain security, approving the funding via the CIA and enlarging an AEC site to accommodate flight testing, Eisenhower stayed out of the specific development work. Such operational work was not in his area of expertise nor would he add value by getting involved. Inevitably, this would only slow the process. He empowered Bissell to operationally manage the development on the understanding that the Washington-insider Bissell would cut through red tape and find ways to enable the technological developments in a timely and cost-effective manner. Equally, Bissell, neither a scientist nor an engineer, empowered Lockheed's Kelly Johnson to deliver a radically different aircraft, and empowered figures such as Land, Baker, and Art Lundhal to develop the tools of photo-intelligence. The fact that so much CIA material on the development of the U-2 remains classified today is testament to the success of the project. Indeed, the U-2 remains in service with the US Air Force. Approval for the initial development project was given on December 1, 1954. The plane flew for the first time on August 8, 1955, and the first Soviet over-flight took place on July 4, 1956.[64] Eisenhower kept a close rein on the U-2, personally signing off all development stages and, more crucially, signing off each mission. However, he gave his subordinates the freedom to create the best possible intelligence-gatherer in the best possible way. Bissell's explanation of the development process is an apt description for why the project would not have worked under air force control and, indeed, one can extrapolate from it much of the reasoning for the delays and duplications that bedeviled the parallel ICBM/IRBM development.

> Lockheed had by then become personified by [Kelly] Johnson and could be trusted not to abuse the government's interest. The CIA, as the procuring organization, was willing to delegate major authority to him and to Lockheed. Thus he and I were able to cut through layers of red tape and reporting procedures that would have slowed the project

> down. We kept our regular monthly progress reports to about five pages, for example; had the same program been developed for the Air Force, it would have required the preparation of a document one inch thick. If the project officer for an Air Force plane decided a change was required, he had to check with Wright Field, a couple of different laboratories, the budget office, the regulations office and so forth—that was one reason it was difficult to receive prompt, clear answers that would not be subject to repeated reviews.
>
> About two years after the U-2 was up and flying, the officer in charge of procurement for the Air Force . . . [inquired] into the CIA's unique ability to support the rapid development and deployment of an aircraft system. . . . [He] expressed surprise the agency used the same procurement office for wings as it did for the engines. The Air Force would never do that.[65]

Bissell's memoir provides extensive detail on the development of the U-2 and its success from 1956 to 1960 in overflying the Soviet Union and its satellite countries and providing intelligence that satisfied the president that so much of the Soviet rhetoric about its growing nuclear missile and bomber fleet threat was merely that: rhetoric. It was always intended by the TCP that the U-2 would be the first in a series of scientific intelligence gathering options and indeed, the A-12 supersonic jet project was signed off by Bissell, acting for Eisenhower, on January 30, 1960.[66] By this time though, Bissell's team were already working on CORONA—another example of hiding in plain sight.

SATELLITE DEVELOPMENT

Following the TCP report, the Air Force announced in March 1955 that it was to develop a reconnaissance satellite. In fact, the TCP report echoed a RAND study, co-sponsored by the CIA from 1954 which recommended that the Air Force should work towards "at the earliest possible date completion and use of an efficient satellite reconnaissance vehicle as a matter of vital strategic interest to the United States."[67] This study had dovetailed with the newly-established Air Force research and development command's initiation of project 409-40, later designated WS117-L. However, this began to gain considerable coverage in the aviation media since its existence was completely unclassified. For someone with a natural urge towards secrecy, that was entirely unacceptable to Eisenhower. In March 1958, as development on the CORONA satellite reconnaissance system neared completion, Land prompted Eisenhower to turn what was ostensibly a white program black. Eisenhower was a master of removing such discussions from the formal record, but a handwritten note from his staff secretary, Goodpaster, records the president asking for Bissell, Allen Dulles, Killian, Secretary of Defense McElroy, his assistant Quarles and U-2 designer Kelly Johnson to meet in the Oval Office to discuss CORO-

NA.[68] According to Bissell, the result of the meeting was that the satellite element of the project should be turned over to the same CIA group, supported by the air force, responsible for the U-2.[69] Such was the secrecy attached to this project, especially as it came just six months after the launch of Sputnik 1, and less than three months after the launch of the first American scientific satellite, that it was said by Bissell that Eisenhower's final approval was handwritten on the back of an envelope.[70] That envelope has not survived, but the ever-efficient Goodpaster drafted a memorandum for record noting that CORONA had been approved.[71] In fact, it was Goodpaster's efficiency and occasional enigmatic diary entries from Kistiakowsky, by then chairman of PSAC and the President's Special Adviser on Science that provided what little White House paper trail exists to track CORONA's progress. Eisenhower is notable by his absence from this record. An entry from Kistiakowsky in his diary dated July 16, 1959, for instance, stated that following the regular NSC meeting: "Bissell asked for a meeting with the President at 10 am Monday regarding CORONA. He asked Killian and I to attend."[72] This paragraph was cut from the published diaries in the mid-1970s, perhaps reflecting the secrecy that still attended the project.[73] The following Monday's entry contained an equally enigmatic sentence—now not mentioning CORONA at all. "With the President, Bissell, Killian and McElroy for an hour. Classified project approved, although the President expressed concern about high costs in immediate future."[74] This project was the extension of CORONA through the summer of 1960 as detailed in Bissell's outline, sent to Goodpaster, as the link with Eisenhower, on July 7, 1959. The extension was for four more CORONA flights "for the purpose of obtaining precise geodetic fixes and for the extension of existing datum planes throughout the Soviet Union."[75] It is rather frustrating to historians that the estimated cost of this extension to CORONA is redacted in the declassified document.

CORONA was not an immediate success in the mold of the U-2. However, Defense Secretary Gates, who succeeded McElroy, noted the following in his "Summary of Progress on the Military Space Projects during June, July and August 1960":

> DISCOVERER XIII and XIV were launched into polar orbit on the 10th and 18th of August respectively. After orbiting the earth for over 26 hours, both capsules were recovered. DISCOVERER XIII was recovered from the sea and DISCOVERER XIV was snatched from the air by an Air Force C-119. These events marked the first time in history man-made objects, which had been in orbit around the earth, were returned and recovered.[76]

There is no little irony that CORONA's first successful mission, hiding in plain sight as "Discoverer," during which the booster launched successfully, the satellite achieved orbit, the cameras worked and ejected their film payload properly and that payload was recovered successfully occurred as shot-down U-2 pilot Francis Gary Powers was being tried in Moscow. The output of that mission alone covered more than one million square miles of Soviet territory—according to Bissell: "more coverage than all the pictures taken of that country during the entire U-2 program."

Eisenhower had played a powerful role in the development of US reconnaissance capability. And contrary to any residual impression of his passivity in office, those who worked directly with him soon gained respect for his critical and sometimes reflective means of information gathering and processing. "I always found Eisenhower business-like and even tempered," wrote Bissell. "He was well-informed, wanting to know everything about the question before him, and an intent and intelligent listener. I never saw him exhibit anger in any discussion of policy."[77] To some extent, Bissell was fortunate. Eisenhower could get angry and cantankerous. But evidence suggests that this was when subordinates brought him problems rather than recommendations or better still, solutions to issues. In swiftly advancing the United States' reconnaissance capability, he scored a significant success and was well served by a line of excellent operators from Killian through Bissell to Johnson, Land, and Baker who applied rare skills to uncommon problems with stunning results.

When it came to the parallel ICBM/IRBM development, the United States was successful, but not with the speed or clarity of purpose displayed in the U-2/A-12/CORONA programs. As discussed before, the issue was quite simply that Eisenhower was not, in the middle of his first term, prepared to challenge the established practices of the Pentagon—even though those practices were bywords for delay and dysfunction. In *Mandate for Change,* Eisenhower wrote:

> It might have been best, had it been feasible, to remove the whole missile program from the hands of the regular military services and to establish another "Manhattan Project" similar to that through which the atomic bomb was developed during World War Two. This scheme would have had the advantage of concentrating the best scientific minds on one set of programs and eliminating duplication and rivalry among the various service activities. However, by the time the urgency of the program became apparent, each of the services had already organized and was using experimental teams of scientists and engineers for missile development. To tear up all of these organizations and to transplant the scientist, engineers and officers already engaged in the business seemed to me, to Secretary Wilson and to my military-scientific advisers to promise more delay than would continuation of existing procedures.[78]

Eisenhower's attempt at simplification and streamlining was carried on the back of Charles Wilson, or "Engine Charlie" as the former president of General Motors and now secretary of defense was known. The Wilson/Eisenhower Reorganization Plan Number 6 of 1953 made some changes at the highest levels of the armed forces, but further efforts at unifying the services further were largely superficial and effectively stymied in Congress after November 1954 when the Democrats wrested control from the Republicans. Laying down the law on missile development would provoke battles with the services and, potentially with congressmen intent on benefitting from the military-industrial pork barrel. It was a battle the executive was not sufficiently strong to face, and saw no need to face while the US nuclear deterrent could be managed effectively by the SAC. Yet even knowing the power of the SAC at the time, there is an underlying sense of self-justification in the president's explanation as to why he did not press harder for a unified approach to missile development.

CONCLUSION

Eisenhower was bound by duty and a desire to serve the United States to the best of his abilities. He expected that same sense in others and could be cutting, dismissive, and cantankerous when he did not see it demonstrated. In character, he was a planner: prepared to plan every possible scenario, even though he knew that in the field, be that field political or military, his plans were never likely to be delivered as they had been imagined on paper. As a leader, he was not a battle general. Until the Louisiana Maneuvers, his main experience had been as a staff support officer, and even in those mock battles, his operational role was subsequently exaggerated by admirers. Even when supreme commander, his key achievements were getting the right people in the field at the right time with the right logistical support to achieve victory.

As president, he believed that it was not his role to get minutely involved in the deep detail of running the country. He set the direction, surrounded himself with the people he trusted to deliver on his ambition and got involved in issues directly only when those in his command could not deliver his intent. He had inherited a country at war, and was perhaps the best prepared White House incumbent to have to deal with the consequences of war. He did so pragmatically, not by rolling back communism, but by finessing a ceasefire and putting energy behind peace talks that were already underway. Of course, he also made quite clear the threat of heavy and sustained bombing of North Korea if it didn't help the peace talks to an acceptable conclusion! In "waging peace" for the ensuing eight years, he adopted a rational, realist stance that enabled the avoidance of war through an overt build-up of a force capable of delivering unthinkable retribution on the Soviets via the principle of massive retaliation. It was a very logical reliance on a military threat solution coupled with a high personal sense of duty in never having to turn to such massive retaliation.

According to his vice president, Richard Nixon: "When it came to making a final decision, he was the coldest, most unemotional and analytical man in the world."[79] That is a reasonable assessment of Eisenhower, and substantially different from Neustadt's assessment which said:

> When he could not work through a set procedure, or when channels failed him, or when his associates quarreled openly, he grew either disheartened or enraged.[80]

The Eisenhower who adopted the New Look defense policy did not appear disheartened even after months of fractious negotiation with the Joint Chiefs of Staff.[81] In fact, the current state of the historiography paints a rather more pragmatic, even ambitious figure. In 2013, Robert Jervis said:

> Eisenhower came in at a very difficult period. The danger of war was seen as very high. The American economy was seen as unstable with danger of recession and inflation. And the US was facing situations that were literally unprecedented.[82]

Eisenhower succeeded in spite of these difficulties. He made difficult but pragmatic decisions. He welcomed expert insight and was prepared to act on it, and he found original and innovative means to deliver the tools necessary for continuing national security without submitting to demands for ever more funding for ever more military projects. In the same session as Jervis, Evan Thomas remarked:

> Eisenhower did struggle with his own military. He was frustrated by it, but he did cut it by 20 per cent. And however painful it was for him, it would have been a lot worse had liberal democrats been in control.[83]

Yet, even these latest revisions to Eisenhower's reputation fail to capture the guile and elegance he demonstrated in turning some of the most vital recommendations of the TCP into action. By the end of 1955, he had a policy in place to develop intercontinental missiles as a highest priority and was close to the launch of the first effective secret reconnaissance of the Soviet Union in the history of the United States. He had initiated the basic tools necessary for national security and to enable feasible space exploration. In 1954–1955, Eisenhower took decisions that made far greater change in intelligence gathering than the United States had ever seen before or arguably since. His direct intervention ensured Bissell and his U-2 team could bypass the bureaucracy of the Air Force and put a spy in the sky over the Soviet Union more than a year before the launch of Sputnik. The flights of the U-2 provided detailed evidence that there was no bomber gap, and would later refute the Democrats' continued claim that there was a missile gap. His New Look defense policy put emphasis on the development of ICBMs and IRBMs based on the knowledge that the nuclear deterrent could be carried effectively by the Strategic Air Command for the remainder of the 1950s, giving time for each of the services to develop missile systems within the constraint of overall falling defense budgets. The Castle-Bravo nuclear test had proven the United States had the means to create much smaller, more powerful nuclear devices that could be carried, in the short-term, by the B-52 bomber, and would not require the kind of huge rocket booster envisaged by either RAND or Von Braun at the end of the 1940s. Having brought together a strong, focused, and impartial team through the TCP, he was confident that the United States was on course to effectively deliver his New Look policy.

Still, his actions do not meet the super-human standards of a true "Hidden Hand" president as envisaged by Greenstein's model. Refusing to take a tougher stand on missile development was a failing and would cost Eisenhower in terms of bipartisan political and, albeit briefly, public support in 1957. However, by the time of the launch of Sputnik, the ICBMs and IRBMs he authorized after the TCP report were close to readiness and the successful series of U-2 Soviet over-flights proved that Soviet bomber and missile developments presented no immediate threat to the United States. The problem was, once Sputnik 1 was launched, he was not prepared to tell the American people why and how he knew it posed no threat. His critics pounced on that failure to communicate immediately after Sputnik, and there is a case for criticism there. But it is clear that the interpretation of NASA (and the US space program) as a direct reaction to Sputnik is quite simply wrong. For Murray and Bly Cox to say:

> NASA was cobbled together by Congress and President Eisenhower. NASA was a reaction to the panic caused by the Soviet Union's launch of Sputnik 1 in October of 1957. . . . To President Eisenhower, this [panic] was all a lot of hysterical poppycock; but he also decided that he couldn't ignore it. The United States was going to have a space program whether or not he wanted it,[84]

is simply not credible. Eisenhower had the basis of a pro-active space policy two years before Sputnik was launched.

NOTES

1. Defense Manpower Data Center figures, as of May 16, 2008, www.koreanwar-educator.org/topics/casualties/pdfs/korea.pdf. Accessed January 4, 2013.
2. DDE, *Mandate for Change—The White House Years 1953–56* (London, 1963), p. 446.
3. Ibid., pp. 446–47.
4. S Dockrill, *Eisenhower's New Look National Security Policy 1953–1961* (London, 1996), pp. 1–40.
5. NSC 162/2 "Report to the National Security Council by the Executive Secretary on Basic National Security Policy," October 30th, 1953, Document 163, Intelligence

Documents, *Foreign Relations of the United States, 1950–55*, Office of the Historian, Department of State, history.state.gov/historicaldocuments/frus1950-55Intel/d163. Accessed September 8, 2014.

6. Dockrill, *New Look*, p. 275.

7. MJ Hogan, *A Cross of Iron, Harry S Truman and the origins of the National Security State 1945–54* (Cambridge, 1998), p. 467.

8. NSC 162/2 "Report to the National Security Council by the Executive Secretary on Basic National Security Policy," October 30th, 1953, Document 163, Intelligence Documents, *Foreign Relations of the United States, 1950–55*, Office of the Historian, Department of State, https://history.state.gov/historicaldocuments/frus1950-55Intel/d163. Accessed September 8, 2014.

9. NSC 68 "Report to the National Security Council by the Executive Secretary on the United States Objectives and Programs for National Security," April 14, 1950, President's Secretary File, Papers as President, Truman Papers, Truman Library, http://www.trumanlibrary.org/whistlestop/study_collections/coldwar/documents/pdf/10-1.pdf. Accessed September 8, 2014.

10. Memorandum by the President to the Secretary of State, Denver, September 8, 1953, Whitman File, Eisenhower Papers as President, DDE Library.

11. S. Everett Gleason, Memorandum of Discussion at the 163rd Meeting of the National Security Council, Thursday, September 24, 1953, drafted by Deputy Executive Secretary Gleason on Sept. 25, Whitman File, Eisenhower Papers as President, DDE Library.

12. NSC 162/2, pp. 19–24, as per note 147. Eisenhower formally approved NSC 162/2 on October 30, 1953.

13. DDE, *Mandate for change*, p. 447.

14. Ibid., p. 449.

15. Ibid., p. 452.

16. Unattributed, "Richard Bissell, an Agency Leader, the People of the CIA," Featured Story Archive, 2012, https://www.cia.gov/news-information/featured-story-archive/2012-featured-story-archive/the-people-of-the-cia-bissell-an-agency-leader.html. Accessed April 8, 2014.

17. Among the many titles that cover the subject well, Neil Sheehan's *A Fiery Peace in a Cold War* (New York, 2010), stands out as providing a comprehensive review of the early US-Soviet arms race through his study of the Air Force's General Bernard Schriever who set up the Air Force's Western Development Division under the Air Research Development Command, delivering both the first US ICBMs, and the Atlas and Titan rockets that enabled many of the early US satellite launches. This research touches on Schriever's work only where it intersects with the actions of Eisenhower. The same can be said for the development of both the U-2 and CORONA reconnaissance satellite. Both of these developments, which were outputs of the NSC 162/2 strategy, are superbly documented in William E. Burrows' *Deep Black*, (London, 1998), while Philip Taubman's *Secret Empire* (New York, 2003) brings a rather more journalistic angle to air and space intelligence gathering in the late 1950s and early 1960s.

18. DDE, "Address to the National Republican Convention," Chicago, July 30, 1960, http://www.presidency.ucsb.edu/ws/?pid=11890. Accessed September 9, 2014.

19. J Harford, *Korolev* (London, 1999). James Harford's scholarly biography of Sergei Korolev, the Soviet rocket program's "Chief Designer" provides an excellent insight into the early philosophy and developments in the Soviet space and missile setup.

20. Taubman, *Secret Empire*, p. 67.

21. Kalic, *Militarization of Space*, p. 24.

22. The Air Force had established a new office of Deputy Chief of Air Staff for Research and Development, to which Project RAND would report, in December 1945. General Curtis LeMay took the role. Three months later, the RAND team was installed in the Douglas works in Santa Monica. Their first report "Preliminary Design of an Experimental World-Circling Spaceship," which was concerned with the potential

design, performance, and possible use of man-made satellites, was published on May 2, 1946, http://www.rand.org/pubs/special_memoranda/SM11827.html. Accessed May 12, 2014.

23. DDE, "Scientific and Technological Resources as Military Assets," Killian Collection, MIT, as referenced in chapter 1, footnote 165.

24. Douglas Aircraft Inc., Santa Monica Plant Engineering Division, "Preliminary Design of an Experimental World-Circling Spaceship," Report No. SM11827, May 2, 1946, p. 1.

25. Ibid., p. 2.

26. S Ambrose, *Eisenhower—Soldier and President* (London, 1990), p. 365.

27. DDE: The President's News Conference, March 24, 1954. Online by Gerhard Peters and John T. Woolley, The American Presidency Project, http://www.presidency.ucsb.edu/ws/?pid=10191. Accessed April 8, 2014.

28. Excerpts from DDE News Conference, March 31, 1954, participants: DDE, Lewis L. Strauss, Chairman of the AEC, and various news correspondents, www.nuclearfiles.org/menu/library/correspondence/eisenhower-dwight/corr_eisenhower_1954-03-31.htm. Accessed April 8, 2014.

29. R Ferrell, ed., *The Diary of James C Hagerty: Eisenhower in Mid-Course, 1954–1955* (Bloomington, IN, 1983), entry for March 31, 1954, pp. 36–37.

30. *Washington Post,* April 1, 1954, Section A-10.

31. The report referred to is NSC-140, which marked the establishment by Truman of the Special Evaluation Subcommittee of the NSC—www.eisenhower.archives.gov/research/online_documents/aerial_intelligence/Chronology.pdf. Accessed April 8, 2014.

32. "Meeting the Threat of Surprise Attack," A Report from the Technological Capabilities Panel to the President, Washington, DC, February 14, 1955, Office of the Historian, US Department of State, https://history.state.gov/historicaldocuments/frus1955-57v19/d9. Accessed September 8, 2014.

33. Killian, *Sputnik,* pp. 67–93.

34. DE Welzenbach, "Science and Technology: Origins of a Directorate" in *Studies in Intelligence* Vol. 30, No. 2 (1986), p. 17.

35. Welzenbach, Science and Technology, p. 13.

36. Executive Order 8807, establishing the Office of Scientific Research and Development, 28 June 1941, unattributed, *The Public Papers and Addresses of Franklin D Roosevelt* (New York, 1941) p. 239.

37. Killian, *Sputnik,* p. 68 and C Peebles, *High Frontier,* Air Force History and Museums Program (Washington, DC, 1997) p. 4.

38. Memorandum for the Secretary of Defense from Donald A Quarles, Assistant Secretary of Defense, 24 May 1954, Killian Papers, MIT.

39. Burrows, *Deep Black,* p. 69.

40. Outlined in "Appendix on Organization and Activities of the TCP in Meeting the Threat of Surprise Attack Report by the Technological Capabilities Panel of the Science Advisory Committee," (Washington, DC, February 14, 1955) Foreign Relations of the United States, 1955–1957, Volume 19, National Security Policy, Document 9, Office of the Historian, Department of State, https://history.state.gov/historicaldocuments/frus1955-57v19/d9. Accessed September 8, 2014.

41. Ibid.

42. Ibid., Section C, p. 9.

43. Ibid.

44. Memorandum of Discussion at the 257th Meeting of the National Security Council, Washington, August 4, 1955, NSC Records, Whitman File, Papers as President, DDE Library.

45. Z Wang, *In Sputnik's Shadow* (New Brunswick, NJ, 2009) p. 51.

46. Burrows, *Deep Black,* p. 70.

47. Ibid., p. 73.

48. In 1955, manufacturing accounted for 30 percent of the US workforce which stood at 65 million workers, with defense being the largest single sector within this just ahead of automobiles. MA Lee, *US Labor Force Trends*, in *Population Bulletin*, Vol 63, no. 2, June 2008, Population Reference Bureau, (Washington, DC). In Southern California alone, almost 300,000 workers were employed in defense activities for the major aerospace and missile contractors by 1955. For a very detailed breakdown of the statistics, see AJ Scott, *Technopolis:* High-Technology Industry and Regional Development in Southern California (Berkeley, 1993).

49. *Collier*'s circulation peaked in 1944 at 2,846,052, but was in decline in the mid-1950s as it was increasingly challenged both by newspaper supplements and the rise of TV. It ceased publication in 1957. However, in 1955, it briefly arrested its decline in readership by publishing a series of articles entitled *Man Will Conquer Space Soon*, https://colliersmagazine.com/about-us. Accessed July 22, 2014. *Aviation Week*, the leading trade title for the aviation and defense industries had a circulation of 62,970 in the United States in 1957, according to the Audit Bureau of Circulations, October, 1957.

50. Document 34, Volume 19, National Security Policy, *Foreign Relations of the United States, 1955–1957*, Officer of the Historian, Department of State, https://history.state.gov/historicaldocuments/frus1955-57v19/d34. Accessed September 8, 2014.

51. VT Ford, oral interviews with J Lonnquest, 1992, subsequently published in JC Lonnquest and DF Winkler, *To Defend and Deter, the Legacy of the US Cold War Missile Program*, USACERL Special Report 97/01, (Washington, DC, 1996).

52. VT Ford, *Twenty-Four Minutes to Checkmate*, unpublished manuscript, Volume 3, part 3, Box 2, Vincent Ford Papers, DDE Library. At times, the manuscript verges on melodrama, and, written after the event from Ford's notes, there is a question of veracity. However, Ford's narrative is supported by the rather more anodyne NSC report of events which brought the ICBM program back to the table on September 7, 1955. Here, the report stated: "The Killian Group recommended that the NSC formally recognize the present Air Force program for the development of an intercontinental ballistic missile 'as a nationally supported effort of highest priority.' . . ." The NSC was briefed on this subject at its July 28 meeting. This was an oral report, and the NSC memorandum of discussion of that meeting merely took note of the fact that this report had been given. (NSC Records, Whitman File, DDE Library). However, the September 7 document also noted: "In the Defense Department the ICBM program is already rated as 'highest priority.' About 180 other projects, however, are said to enjoy this same rating. Moreover, it is understood that, even with this rating, the ICBM program has been seriously delayed by the existing procedures for clearance and approval." R. Bowie, Intercontinental Ballistic Missile Program, Memorandum from the Director of the Policy Planning Staff to the Acting Secretary of State, September 7, 1955, (Washington, DC) Document 33, National Security Policy, FRUS 1955–57.

53. Ford, *Twenty-Four Minutes*, p. 292.

54. Ibid., p. 293.

55. Richard M Bissell biography, CIA Archive, www.cia.gov/news-information/featured-story-archive/2012-featured-story-archive/the-people-of-the-cia-bissell-an-agency-leader.html. Accessed April 8, 2014.

56. Interview with Richard M. Bissell Jr., conducted by Theodore A Wilson and Richard D McKinzie, July 9, 1971, p. 67, Oral History Series, Harry S Truman Presidential Library, www.trumanlibrary.org/oralhist/bissellr.htm. Accessed September 9, 2014.

57. Killian, *Sputnik*, p. 82.

58. Bissell, Oral History, p. 67.

59. AJ Goodpaster, "Memorandum for Record," The White House, April 21, 1958, Box 14, Intelligence Matters, Office of the Staff Secretary, DDE Library.

60. Killian, *Sputnik*, p. 82.

61. Bissell, *Cold Warrior* (New Haven, CT, 1996), p. 95; Burrows, *Deep Black*, p. 74.

62. Bissell, *Cold Warrior*, p. 96. Given the intense secrecy around the project, Bissell's recollection in his memoir is the best (and perhaps only) record of the meeting available to scholars.
63. Ibid., pp. 102–3.
64. Ibid., p. 92.
65. Ibid., p. 99.
66. Burrows, *Deep Black*, p. 158.
67. RAND/CIA, "Project Feed Back Study Summary Report," March 1, 1954, www.rand.org/history. Accessed April 9, 2014.
68. Handwritten notes of General Andrew Goodpaster, March 24, 1958, Box 14, Intelligence Matters, Office of the Staff Secretary, DDE Files, DDE Library.
69. Bissell, *Cold Warrior*, p. 135.
70. Peebles, *High Frontier*, p. 12.
71. Memorandum for Record by General Andrew Goodpaster, April 21, 1958, Box 15, Intelligence Matters, Office of the Staff Secretary, DDE Files, DDE Library.
72. Kistiakowsky Diary, July 16, 1959, pre-publication record, G Kistiakowsky Diary File, Organization and Early History of NASA Collection, DDE Library.
73. GB Kistiakowsky, *A Scientist at the White House*, (Cambridge, MA, 1976), p. 14.
74. Kistiakowsky, *Diary, pre-published transcript*, July 20, 1959.
75. CIA/Bissell Memorandum to Goodpaster, "Subject: Proposed Supplement to CORONA Project," July 7, 1959, Box 14, Intelligence Matters, Office of the Staff Secretary, DDE Library.
76. TS Gates Memorandum: "Summary of Progress on the Military Space Projects during June, July and August 1960," October 24, 1960, DoD Box 9, Staff Secretary Files, DDE Library.
77. Bissell, *Cold Warrior*, p. 114.
78. DDE, *Mandate for Change*, p. 456.
79. Nixon, *Six Crises*, pp. 158–59.
80. Neustadt, *Presidential Power*, p. 137.
81. See Hogan, "The Iron Cross: Solvency, Security and the Eisenhower Transition" in *A Cross of Iron* for a forensic reconstruction of the budgeting process around defense policy from 1953–1955.
82. R Jervis, conference comments at *Waging Peace, Eisenhower and Foreign Policy*, Session 4, "Ike Reconsidered, Lessons from the Eisenhower Legacy for the 21st Century," March 8, 2013, Roosevelt House Public Policy Institute, New York.
83. Evan Thomas in Session 4 at "Ike Reconsidered." March 2013, New York.
84. Murray and Bly Cox, *Apollo*, pp. 23–24.

THREE

The Sputnik Season

Despite the 1980s being the decade of Eisenhower revisionism, Murray and Bly Cox, writing in 1989, still described NASA in terms of a "reaction to the panic caused by . . . Sputnik 1."[1]

They added:

> The public furor surrounding these events [the successful launches of Sputnik 1 and Sputnik 2] had been immense. . . . In their humiliation, Americans lashed out at a variety of targets—the educational system, the military, Eisenhower's golf, American consumerism."[2]

This interpretation has slipped into common usage, and continues to project Eisenhower as weak and passive in the face of Soviet success, acting only as a response to the pressures inflicted by the American public. In 2007, Tindall and Shi wrote:

> On October 4th, 1957 the Soviets launched the first satellite called *Sputnik*. Americans, until then complacent about their technical superiority, panicked. . . . The Soviet success with Sputnik led to efforts in the United States to increase defense spending, offer NATO allies intermediate-range-ballistic missiles pending development of long-range intercontinental ballistic missiles (ICBMS), set up a new agency to co-ordinate space efforts and establish a crash program in scientific education and military research.[3]

The observation that "Americans panicked" is an element of received wisdom that features in the arguments of many scholars who interpret the events in the Fall of 1957. It is often driven by noting the newspaper headlines and television sound bites in the immediate aftermath of Sputnik 1. Matthew Brzezinski, in *Red Moon Rising* for instance, painted an evocative picture.

> The warning [of impending Soviet superiority] was echoed by thousands of media outlets, big and small, conservative and liberal, in radio and television, magazines and newspapers. . . .
>
> A strange sense of disconnection gripped the public discourse. The more the Administration told America not to worry, the louder the media beat the doomsday drums.[4]

Even among those most associated with Eisenhower scholarship in recent years such as Launius, the interpretation of public hysteria greeting Sputnik and Eisenhower only moving forward on space issues reluctantly under outside pressure still holds sway.

> Eisenhower had refused to fall prey to public hysteria over the Sputnik launches in 1957, and set in place, only with some reluctance, NASA as an independent Executive branch agency in 1958. . . .
>
> In the crisis over Sputnik, Ike had felt intense pressure from an alliance of diverse interests to establish a cabinet-level federal entity, something he always thought unnecessarily expensive, and once created, almost impossible to dismantle, to carry out a visible program of space exploration.[5]

Such views simply do not square with the evidence now available. So it is high time to revisit the fall of 1957 to reassess the evidence of what actually happened by focusing on two elements that remain cornerstones to the orthodox approach to the impact of the "Sputnik Crisis," namely the reaction of the media and other interested actors to the Soviet launch and Eisenhower's actions both in the immediate aftermath of the first Sputnik launch and also in the period from his first news conference following the event on October 9, 1957, to the day of his stroke on November 25, 1957.

Far from being cowed by the Soviet success and the reaction of his political opponents to it, Eisenhower actually went on the offensive to the extent that by November 14, less than two weeks after the launch of Sputnik 2, he had regained the initiative with the media and, more so, had taken the opportunity to drive forward his own national security agenda as discussed in the previous chapter. There is another clear distinction to be drawn that is hardly discussed in the mainstream Sputnik discourse. That is, the very distinct difference between the American response to Sputnik 1 and Sputnik 2. These separate events are too often conflated into a single narrative.[6]

The most important point to make is that Eisenhower's reaction was logical. Following the launch of Sputnik 1, there was no clear and present danger to the United States, and therefore no need for any heightened response. However, this lack of a visible response created a vacuum and a damaging sense of passivity, even if this was not actually the case. The launch of Sputnik 2 was an opportunity to correct the misperception. This time around, Eisenhower's actions were more obvious, and also more clearly attuned to the public mood. However, the public, and the media shaping its opinion, had already begun to move on by that time. Sputnik 2 did not inspire fear and awe and the reaction to it was very different from what occurred the month before.

THE CONTEXT FOR SPUTNIK

Sputnik was an outcome of the the International Geophysical Year (IGY). But the IGY itself was a misnomer. A successor to the two previous International Polar Years (1882 and 1932) it was proposed by the American International Council of Scientific Unions in 1952. The plan was to study the whole earth over the next period of maximum solar activity. It was not actually a year, but the eighteen months from the start of 1957 to the middle of 1958. Given that this was a whole-earth study, and also included projects observing and measuring activity in the upper atmosphere, the study became a geophysical rather than polar year, and its international element was cemented in 1953 when sixty-four countries signed up to studies within the umbrella program. Led by James S Lay, the NSC planning board, at the request of the Department of Defense, produced NSC 5520 the "Draft Statement of Policy on a U.S. Scientific Satellite Program,"[7] which recommended the creation of a scientific satellite program as part of the International Geophysical Year as well as the development of satellites for reconnaissance purposes. Largely, this was an outcome from the TCP, which wrapped the scientific satellite project into the overall national security project. Based on this report, the National Security Council approved the IGY small scientific satellites program on May 26, 1955.[8] However, it was not until July 28, 1955, that Eisenhower's press secretary, James Hagerty, made the news public in a briefing to reporters at the White House. The formal statement was dated July 29, since the information could not be made public until it had been shared with Marcel Nicolet, the secretary of the Brussels-based Comité Spéciale de l'Année Géophysique Internationale.[9] Hagerty emphasized that the satellite program was intended to be the US contribution to the IGY and that the scientific data drawn from the study was to benefit scientists of all nations.[10]

Meanwhile in Moscow, the USSR had responded to the call for IGY projects with its own declaration that it intended to put a satellite into orbit during this special period of scientific study. According to his biographer, James Harford, the chief designer of the Soviet rocket program, S. P. Korolev, had first proposed using the R7 rocket launcher to place a satellite in orbit to the Central Committee of the Communist Party towards the end of 1953.[11] Soon after Hagerty's announcement of the US intent to develop a US scientific satellite as part of the nation's IGY initiative, the chairman of the Soviet Academy of Sciences Commission on

Interplanetary Communications, Leonid Sedov, seemed to confirm the Soviet commitment when he told a group visiting the Soviet Embassy during the Sixth Congress of the International Astronautical Federation meeting in Copenhagen: "From a technical point of view, it is possible to create a satellite. . . . The realization of the Soviet project can be expected in the comparatively near future."[12] Harford suspected that Sedov was himself speculating since the Soviet Council of Ministers did not issue a decree authorizing the development of any IGY satellite until January 1956.[13] There was more than a certain irony in the fact that US media had been predicting a satellite launch since 1955 when the US Vanguard program was announced. In fact, Alton Blakeslee at the Associated Press gave a very accurate prediction of the likelihood of a satellite in 1957 in a syndicated article that was run by, among many other newspapers, the *Courier News,* in Blytheville, Arkansas, right at the start of 1957. The paper reported:

> MORE STRIDES IN CONQUEST OF SPACE COMING IN 1957
> With roaring rockets, man begins his conquest of space in 1957. Scientists will continue test-firing the rockets designed to bang a little artificial moon into outer space. Whizzing around the earth every 90 minutes, it will be the first man-made messenger exploring the puzzles and hazards of our next great frontier—space travel.
>
> The first little "moon," about the size of a basketball, may even be launched within the next 12 months. You may even be able to see it as a faint, fast-moving dot of light low in the sky at dusk or morning.[14]

As a prediction, it was startlingly accurate. But it made one crucial flawed assumption. That was that the "artificial moon" would be American. The article went on to talk about the wide US efforts in the IGY, and made no mention of the Soviets at all. By May, Blakeslee had changed his tune and wrote:

> But the Russians could do it first. For the Russians are also planning to shoot man-made moons around the earth. . . . No-one outside Russia knows if they're ahead of us, no single detail has been announced of their progress.[15]

However, stories reflecting US IGY projects appeared in the papers almost every day in 1957, and by the end of August US readers had been led to understand that their first satellite launch was just months away. Yet the Soviets had been quite open about their plans to launch a satellite, and after Vladimir Kotelnikov of the USSR Academy of Sciences presented a paper at a conference in Colorado on August 27 stating that the Soviet satellites were ready for launch, the US press picked up the story and ran it across the newspapers of America.[16] As an example, on August 31, the *Corsicana Daily Sun* in the Texas heartland ran a down-page, page one article headlined: "Manned Satellites Hinted by Moscow." The International News Service syndicated piece stated that in the wake of Kotelnikov's presentation, Radio Moscow had hinted that the Soviet Union planned, as part of their IGY effort, to send much heavier satellites into space than their US counterparts, satellites that "might be hurled into space with humans on board." The article also noted the news that the Soviet Union had announced the successful testing of an ICBM.[17]

Yet, still the US public and legislators chose not to believe that the Soviet Union would be first to launch a satellite. In Congress, there was a strong belief that the USSR did not possess the technology to launch and orbit a satellite. During a Senate Appropriations Subcommittee hearing in 1957 while Werner Von Braun was giving testimony on the possibility of the Soviet Union launching a satellite, Senator Allen Ellender, a Louisiana Democrat questioned whether Von Braun was "out of his mind." It transpired that Ellender had just returned from a congressional junket to Moscow where having seen the state of Soviet engineering, he was convinced that Von Braun's view that the Soviet Union had the capability to put a satellite into orbit was totally wrong.[18]

But in the days before Sputnik was launched, the newspapers were again reporting that a US "first" was not guaranteed. At the end of September, there were reports that the Soviets were talking seriously about an impending launch. For instance, on September 30, 1957, the International News Service syndicated an article headlined: "Russ Scientists to Bare Space Satellite Plans."[19] It reported:

> Three top Soviet scientists arrived in Washington yesterday to give details of Russia's plans for launching earth-circling satellites. . . . The International Geophysical Year satellite and rocket conference . . . program calls for the Russians to lead off with a "technical description of the USSR satellite vehicle and launching program" details still a big secret to western scientists and IGY officials. . . . The Russians will be followed into the spotlight by Dr. John P Hagen of the US Naval Research Laboratory who has been allotted 20 minutes to describe the US program.
>
> Thus, the stage is set for the latest tally on which nation is ahead in the race to get earth circling satellites into space first. The satellite vehicles are, of course, for scientific research purposes and all officials disavow any national competition.[20]

There was no reason that the launch of Sputnik 1 should have been a surprise—especially to journalists and media outlets which had been directly reporting the Soviet IGY plans.

The headlines reported in almost every book and article covering the Sputnik period are accurate. But in this next section, we will discover to what degree they tell the full story.

A SHORT-LIVED MEDIA CRISIS

Following the launch of Sputnik 1, the world's first artificial satellite, on October 4, 1957, the *Chicago Daily Tribune* warned, rather alarmingly, that if the Soviets "could deliver a 184-pound 'moon' into a predetermined pattern 560 miles out into space, the day is not far distant when they could deliver a death-dealing warhead onto a predetermined target almost anywhere on the earth's surface."[21] Meanwhile, the *New York Times* screamed with a rarely-used three deck headline: "Soviet Fires Earth Satellite into Space: It Is Circling the Globe at 18,000 mph: Sphere Tracked in 4 Crossings over US."[22] The media response was largely misinformed extrapolation, but it captured the front pages across the United States, reaching communities large and small. A day after the launch, on October 5, the *Mansfield News-Journal,* in Mansfield, Ohio, fifty miles north of Columbus led with the headline: "Red-Made satellite flashes across the US" and posed the question "First step to moon landing?"[23] The story was the standard United Press syndicated material, supplemented by first-person summaries of seven Columbus, Ohio "moon watchers" who had stayed up all night attempting to tune in to Sputnik's radio signal, and view the satellite via telescope.[24] The following day in California, the *Sun-Telegram* in San Bernardino chose to focus on the political reaction to the launch. Its headline stated: "Soviet Victory Sparks Cry In Congress," and the story, this time drawn from the *New York Herald* News Service, covered the Democrats' demand for an inquiry into how the administration had let the Soviets win what it characterized as the race into space.[25]

Despite the lack of any concrete evidence that the satellite posed any threat to US national security, there was a perception that it did. The perception was that the launch had undermined a central certainty of the capitalist world. Until now, the United States assumed it was technologically superior to the Soviet system (as well as socially, culturally, and politically). At a stroke, the nation's definition of itself as the world leader was rocked—at least from a media standpoint. It is claimed that the real effect of Sputnik was to alter American, and the Free World's, perception of itself, its strength, and its supposed superiority. Even if the launch delivered little in terms of scientific capability, it changed the perception of Soviet capability and by implication, the Soviet threat. However, a careful analysis of those media outlets that were most alarmist shows how transient any sense of panic actually was.

The *New York Times* offers a reasonable case study. By October 10, less than a week after the launch, the tone of the reporting had changed. The third most prominent article on page one of the paper was headlined: "Moscow predicts satellite will stay aloft for long term."[26] By October 20, there were still twenty-nine references to Sputnik in the paper, but no fewer than seven were adverts, including one for an apartment to rent which, like Sputnik, was, according to the vendor, "No pie in the sky deal."[27] By November 1, Sputnik had moved entirely from the news pages to the advertisements. Indeed, the book review pages carried a display advertisement for Max Shilman's "Rally Round the Flag Boys." The advertorial copy stated: "You won't find Sputnik in these hilarious pages, but there are guided missiles galore, misguided marriages and more laughs per page than any book on satellites!"[28] This does not reflect the orthodox image of a nation in panic.

Indeed, not all of the US media reacted in the same way to the Sputnik launch. In 1957, as today, there were no truly national newspapers serving the United States. The *New York Times, Washington Post*, and, to a lesser extent, others such as the *Los Angeles Times* were widely read across the country and saw themselves as the media leaders in setting the opinion of the American public. However, newspapers (in their final years as the prime news medium in the land) were stubbornly parochial; heavily focused on their local or, at best, regional markets and always inclined towards the local market impact of any story. While the New York headline screamed, the *Milwaukee Sentinel* led its October 5 edition with: *Today We Make History*.[29] Sputnik was nowhere to be seen. Instead, the item gripping the attention of the people of Wisconsin was the first-ever World Series baseball game to be played in the city. For the record, the Braves lost game three to the New York Yankees, but recovered to take the series 4–3, thus becoming the first-ever relocation team to take the pennant. Baseball relegated Sputnik to page five of the paper. Indeed, in many local markets, the World Series was seen as significantly more important than the orbiting satellite.

However, many regional newspapers, and also both cinema newsreels and the still entertainment-led TV news networks, took their news from the major national news agencies: Associated Press, International News Services and United Press, part of Scripps-Howard. The output of these agencies ensured that the Sputnik story maintained energy even beyond the initial shock. On October 9, Scripps-Howard syndicated a story across its newspapers and affiliates written by Jim G Lucas and Dickson Preston. "Interservice Fussing Helps Reds Win Satellite Race" attacked "belated claims that the US was never in a race," by using a memorandum from the court martial of John C Nickerson of the ABMA which claimed that the army could have put a satellite in space a year before Sputnik.[30] The pejorative tone of the piece was clear. It did not use named sources nor directly accuse Eisenhower of failings, but was critical of the administration by implication. It concluded:

> Meanwhile the Air Force also was developing an IRBM missile—known as Thor—which today would appear able to do as good a job as Jupiter. But from somewhere came a decision the project would go to the Navy. At the time the Navy had no big missile project of its own. It revived a cancelled missile, the Viking, and decided to make it do the

> job. As Navy spokesmen later told Congress, it considered Vanguard a "bargain basement" project anyway.[31]

These agencies were fierce rivals, each aiming to extend their syndication in a crowded marketplace that was in sharp decline as television began to seriously encroach on its preeminence as a news medium.[32] INS was founded by William Randolph Hearst, twice elected as a Democrat to the House of Representatives. Even after his death, its political allegiance was to the Democrat cause, not least through Hearst newspapers such as the *San Francisco Examiner*.[33] According to his biographer, Scripps was more interested in selling newspapers than in politics, and his heirs used Sputnik 1 and Sputnik 2 as an opportunity to sell as many newspapers and syndicated stories as possible.[34] The Associated Press was (and remains) a not-for-profit co-operative, but even in its role of syndicating coverage from such titles as the *Washington Post* and *New York Times*, it contributed to the flurry of interest in the Soviet achievement.

The other aspect of the Sputnik story that kept it in the news was the local, parochial angle. When the initial surprise at the Soviet launch subsided, newspapers sought to keep the story alive by focusing on interactions in the local community, whether that was political comment, or, as was more often the case, a local citizen tracking the satellite's orbit. In Alabama, the *Athens Limestone Democrat* featured an Athens college professor, James Gillespie, who "picked up the 'beep-beep' of Russia's earth satellite on his 'ham radio' outfit."[35] While repeating the main details of the satellite's launch and orbit, the article focused on Gillespie's history as a radio "ham," and noted he had been born in China. Just four days after the launch, he, rather than the satellite, was the subject of the story. The Sputnik story was not the lead article for the paper that week. That honor went to a story on how the local agricultural community had dealt with severe local storm damage, and was headlined: "Farmers Have Managed to Save 15,032 Bales Cotton."

It is true that the impact of Sputnik 1's launch initially went far beyond its scientific value. It was not seen nor reported simply as a scientific triumph other than by the TASS news agency in *Pravda* and on Radio Moscow.[36] Instead, it was reported as a direct threat to the United States' national security and by implication, the security of the non-communist world. The *Chicago Daily Tribune's* hyperbole: "Death dealing warhead" and "capable of hitting any chosen target anywhere in the world"[37] shows a level of language that was intent, consciously or not, on sowing fear in a society that had only recently come to terms with the Soviets as a thermonuclear power.

It is ironic to note that even in Moscow the initial interest in the launch of Sputnik 1 did not capture the significance of the launch and was distinctly low-key. The Soviet public first learned of the "scientific experiment conducted at such a high altitude" through a down-page article on the front of *Pravda*. The issue that day was led by Marshal Zhukov's visit to Yugoslavia, and the Sputnik announcement was initially made with no hype, little triumphalism and, under just the terse banner "TASS announcement." The article stressed the peaceful and scientific nature of the mission and the authorities clearly had not yet seen the propaganda possibilities that the launch offered. The authorization to proceed with a Soviet satellite project had come from the presidium of the Academy of Sciences as early as August 30, 1955.[38] As the proposal was not for a major weapons project, it received lower priority status and this did not require approval from the party leadership. While the Soviets had stated more than a month before that they were ready to launch a satellite, the actual details of the launch were shrouded in secrecy, with confirmation of the satellite appearing from Moscow only when Sputnik was safely in orbit broadcasting to the world.[39] Significantly, while the *Pravda* report closed with a generic socialist homily, there was no specific political statement; no message from Khrushchev putting down Eisenhower or boasting about the technological achievement from the world's leading socialist economy. This supports the assertion that neither he nor the politburo saw the political and propaganda significance of this scientific achievement at this stage. According to his son Sergei, Nikita Khrushchev was in Kiev when he learned of the launch and was certainly not waiting expectantly for any preplanned outcome.[40] The reaction in the West over the coming days delighted him; it highlighted the significance of the successful launch and gave him a useful propaganda tool for foreign policy. For Khrushchev, from this night on, the space program was never about space exploration. It was a bold display of military might meant to match, and indeed top, America's own frequent displays of firepower.

But the administration was quite aware that the Soviet satellite was likely to be launched first. Overflights of the Tyruratam launch site by a U-2 aircraft in August 1957 had alerted the White House that the Soviets were preparing to launch a rocket, so there was no reason for the government to respond to what was not an unexpected event.[41] But to alert the media from official US sources that a launch was imminent would also have been an alert that the United States had the capability to spy on the Soviet launch site. This would compromise future U-2 operations. Thus, Eisenhower silently faced a dilemma. He could not respond to the short-lived media hyperbole, but for reasons of national security, could not say why.

THE EISENHOWER REACTION

After an early misstep that had much more to do with managing national prestige than with national security, Eisenhower used Sputnik as an opportunity to reinforce his national security strategy and add further weight to the policy he had sanctioned after the TCP report in 1955. Following the launch of Sputnik 1, however, critics have latched on to his silence in the period from October 4 until his news conference on October 9 to present a picture of weak indecisiveness. Immediately following the launch, Eisenhower was neither as nimble nor adept as Greenstein's "Hidden Hand" model would lead one to expect. In fact, critics, including Schlesinger, have given this as a prime example of why the Greenstein model does not work. However, such criticism is unfair both on Greenstein and on Eisenhower. Where it is fair to criticize Eisenhower is in his immediate management of the post-Sputnik message. Essentially, he did not act quickly enough to reach the American public directly and reassure them that the United States was not threatened in any way by Sputnik and nor had the country fallen behind the Soviets in any tangible way. Throughout his presidency up until the Sputnik launch, the operation of the media within the United States had fitted well into the elite model of media influence described by Walter Lippman in 1922.[42] Lippmann asserted that the public is dependent on the media for opinion forming, and the elite model states that the media are controlled by small numbers of people in the political elite, think tanks, and representatives of large business interests. What Sputnik catalyzed was a distinct split within the elite. Politicians such as Lyndon Johnson, defense industry lobbyists and Pentagon budget holders all saw Sputnik as an opportunity to advance their own agendas, and thus operated at odds with the president. A distinction must be drawn here with other members of the elite such as the boards of the Hearst and Scripps news organizations. Their motivation was simply to sell more newspapers and syndicated news stories. For Johnson, here was a high profile issue that could provide a campaign platform for the 1960 presidential race.[43] For corporations including Convair, Lockheed, Martin, RCA, and Boeing, the prospect of the United States competing with the Soviets in space offered the potential for lucrative contracts that would offset any future downturn in their military aircraft business caused by the New Look defense policy's move towards delivering the nuclear threat via ICBMs instead of conventional aircraft.[44] And for the still-competing armed services, there was the opportunity to lobby for primacy in meeting US national security needs for defense against Soviet-launched nuclear missiles.[45] Each faction of the elite moved quickly to publicly air their cause, using the tools open to them both in the daily newspapers hungry for stories that would sell more copies, and the national and trade magazines which depended on advertising sales (often to defense companies) for revenue.[46] This "noise" generated in the days and weeks immediately following Sputnik's launch now sits at the core of the standard retelling. Yet, when one steps back and looks at it afresh, it was extremely short-lived, and largely manipulated by elite actors with something to gain financially, militarily or politically.

The opportunity for the media reaction came the morning of October 5 when James Hagerty informed Eisenhower of the launch. Eisenhower's immediate decision to stay in Gettysburg and focus on his golf game while relegating the initial White House response to Hagerty and Foster Dulles was a poor decision. This was one of the very few occasions across Eisenhower's eight years in the White House when he was poorly advised by Hagerty. The headlines in the major city newspapers the day after Sputnik's launch should have alerted his White House advisers, and particularly the seasoned newsman Hagerty, to the fact that Sputnik was an issue that demanded the direct and reassuring intervention of the president immediately. However confident Eisenhower was that the Soviet satellite was little more than an irrelevance, by staying in Gettysburg and failing to be seen to treat it seriously, he enabled ill-informed sections of the media, fueled by speculative agents looking for political or contractual gain to fill the void and make the job of reassuring the wider American public harder than it should have been.

However, the effect of Eisenhower and Hagerty's media misstep lasted no longer than October 9 when Eisenhower next faced the media in his news conference. While the notes of Hagerty's October 5 daily news conference are all about Sputnik, by Monday, October 7, the interest had already begun to cool, and Hagerty had to deal with just two Sputnik-related questions.[47] There is a case for saying that Eisenhower was right to keep his counsel and let the heat of the immediate media response burn out quickly.

Eisenhower had no more effective supporting operator through his presidency than Hagerty. The former *New York Times* newsman had been on Governor Thomas Dewey's staff from 1943, acting as press spokesman and adviser for Dewey's unsuccessful 1944 and 1948 runs for the presidency. It was this role he undertook in the Eisenhower campaign in 1952, and its success led to his appointment as press secretary. What set him apart from press secretaries both before and since was his strikingly close relationship with Eisenhower. According to Robert Ferrell, who edited Hagerty's political diary as well as the *Eisenhower Diaries*, the bond between the president and his appointee was "an almost father-and-son relationship."[48] The president was instinctively drawn to Hagerty and treated him as a friend and confidant. In fact, Ferrell regarded Hagerty as Eisenhower's "single close friend" in the White House. Hagerty was part of Eisenhower's inner circle for policy making. Ike respected his political judgment and leaned on him to shape policy to ensure it not only appealed to the nation, but in particular, that it satisfied the weak alliance of right-wing and centrist Republicans in Congress.

This may well have been because Eisenhower had not come from that party-political background. While he had a strong political instinct, he relied on those with a direct background in party affairs to advise him on the nuances of Republican Party and congressional thought and likely reaction. Hagerty also had a significantly wider remit than simply media relations. He often chaired the White House staff meetings and sat in (and contributed to) almost every important White House meeting, either working directly with the president, or representing his views and wishes. Like Killian, he instinctively knew what Eisenhower's view would be on an issue, and as such, was entirely trusted and highly valued by the president to fulfill his wishes.

In weighing up whether Hagerty's advice for the president was wrong, had he simply been the president's media expert, and not privy to a deep knowledge of the issues and workings of Eisenhower and his inner circle (and, indeed, be a member of that circle), he might have recognized the potential problems that Eisenhower's lack of immediate overt leadership on the issue would bring. It was an occasion when knowing too much about national security issues meant that the media adviser was not able to advise effectively on how best to manage the post-Sputnik message.

LOW KEY

Eisenhower's actions following the launch of Sputnik 1 were very low-key. This is apparent from the diary entries of his private secretary, Ann Whitman. Whitman was a meticulous diarist and kept a running account of the president's movements, meetings and visitors, along with personal asides on the mood of the Oval Office and the key participants in executive government. Her entry for October 4, 1957, reads:

> In office for about an hour. Many signatures. Saw General Cutler on approval of NSC record of action. Tom and Jim had various problems in connection with the Queen's visit.
>
> Flew to Gettysburg for golf with George Allen.[49]

There is no sense in the entry of any consideration about the launch of a Soviet satellite being front of mind, nor fore-knowledge of what would take place that night. If anything is conveyed, it is the sense that the upcoming state visit from Queen Elizabeth II was the most pressing issue in the minds of the administration. Certainly the coolness of reaction from the White House to the Soviet satellite is reflected in the fact that there are no diary entries at all on October 5, 6, or 7. Eisenhower stayed in Gettysburg, playing golf, dining with friends, and, apparently, thinking about the defense reorganization talks he planned to have with Secretary Wilson early in the coming week. The diary picks up on October 8 when Whitman records: "The President in Office at eight o'clock to discuss Russian satellite with officials of the National Science Foundation. General Goodpaster will prepare notes." While this was the first meeting of the day, Whitman also records that the president was heavily involved in other business, notably meeting with Senator Smith of New Jersey; the finance minister of India together with the Indian ambassador and Sir Brian Robertson, who had served with the president in the North African campaign in World War II. The conclusion that can be drawn here is that Eisenhower still had not attached great significance to the Sputnik launch, but was aware he needed to confirm his own low key reaction by talking the issue through with his science advisers.

At this stage, Eisenhower understood better than most the real situation the Soviets were in, but could reveal little of this without undermining US surveillance efforts. He did not comment officially on the launch until October 9 when he issued a statement prior to his White House news conference, congratulating the Soviets on their achievement.[50] This statement had been carefully constructed in conjunction with Hagerty and Chief of Staff Sherman Adams. But the president had consulted his scientists before issuing the statement. In his memoir, Killian noted that meeting of October 8 when Eisenhower met with Detlev Bronk, president of the National Academy of Sciences, who, along with critiquing the draft statement: "told (Eisenhower) of (William) Golden's original proposal that there should be a full-time Science Adviser to the President supported by an advisory committee of eminent scientists, both located within the White House."[51] Bronk's own account, written some years after the event, so taken with a little caution, noted his response to a question from Sherman Adams as to whether Sputnik called for any alterations to the existing US research and development program, particularly for missiles. Bronk responded: "I see no reason to change our programs. We should not constantly change our programs in response to every action by the Soviets."[52]

Bronk's response is significant for two reasons. First, here was a senior scientist who was aware of the existing programs in hand and advised the president there was no need to change track. Second, it is clear that Eisenhower was already planning for the future, not least in considering upgrading the role of science adviser, and bringing it into the White House. Eisenhower certainly did note Bronk's suggestion and at his urging, sought out the Nobel laureate and eminent physicist, Isidor I. Rabi, chairman of the ODM Science Advisory Committee. Even before Eisenhower formally met the country's leading scientists on October 15, the way was being paved for a high-powered adviser to join the White House team.

Killian fitted the mold for the first presidential science adviser perfectly. He had already proven adept at managing the needs of the president and the abilities of scientists, military leaders, and Washington bureaucrats alike through his shrewd leadership of the TCP. While delivering a final report of depth and clarity, Killian had also produced the result that

Eisenhower wanted, giving him further justification to pursue the New Look defense policy, ultimately at the expense of the expansionist SAC. Killian intuitively understood the president's needs—and was far better placed than most in the scientific community to deliver on those needs. Bringing Killian into the White House inner circle, though it happened in the wake of Sputnik, was much more the fulfillment of a plan that had emerged from the TCP project three years before.

Throughout the post-Sputnik period Eisenhower never lost his composure but his outward calmness concealed a hidden motive. While Eisenhower and the US intelligence community had been evaluating proposals for an orbiting military reconnaissance satellite, they had also been grappling with the political ramifications of likely Soviet reaction to overflights of its territory. The launch of Sputnik effectively ended those concerns, allowing the United States to pursue a policy of space as an "open platform," establishing that national boundaries did not extend into space. Don Quarles, Eisenhower's assistant secretary of defense, commented to the president on October 8 that the Soviets: "have, in fact, done us a good turn, unintentionally, in establishing the concept of freedom of international space," a principle which the Soviets could not now refute. McDougall argued that the administration actually allowed the Soviets to launch a satellite first in order to secure their "open skies" objective.[53] Quarles' comment refutes this interpretation, and suggests that McDougall was stretching the argument too far.

Quarles' comments are captured in a "Memorandum of Conference with the President" drafted on October 9, 1957, by Eisenhower's staff secretary, Goodpaster.[54] It records the meeting referred to by Ann Whitman that the president held the previous day with Quarles; Alan Waterman, director of the National Science Foundation; several of his colleagues; and a number of White House advisers, including Adams and Hagerty. It was the meeting prior to Eisenhower's visit from Bronk. There is no sense of tension or urgency in the note. Goodpaster records a conversation in which Quarles stated that a Redstone rocket "had it been used, could have orbited a satellite a year or more ago. The Science Advisory Committee had felt, however, that it was better to have the earth satellite proceed separately from military development." Noting that the president stated that when this information reached Congress they were bound to ask why this had not happened, Goodpaster paraphrased Eisenhower adding: "timing was never given too much importance in our program, which was tied to the IGY." The tone of the whole memorandum is low-key and relaxed and indeed the group appears to have discussed a response to the Soviet launch. Goodpaster wrote: "The President thought that to make a sudden shift in our approach now would be to belie the attitude we have had all along." According to Goodpaster, Adams had summed up that attitude when he recalled that Dr. Pusey had said that we had never thought of this as a crash program, as the Russians apparently did.[55]

In the popular interpretation of the Sputnik Season, there is a sense that the USSR's action caused first panic and then inaction in the administration. The evidence drawn from the meetings that took place between Eisenhower, his appointees in the DoD, and his advisers at the National Science Foundation suggest otherwise. Undoubtedly there were meetings to assess if Sputnik posed any threat. However, these swiftly concluded that the Soviet capability was known, there was no threat to national security, and Eisenhower's defense strategy was not going to change. He already had a path marked out for progress in space and this was not fundamentally changed by Sputnik.

MEETING THE NATION'S MEDIA

At this stage, it is possible to surmise that Eisenhower felt the storm over Sputnik 1 was dying down and that the voices of anger and outrage would be calmed by the president's complete unwillingness to engage with criticism of his policies and the polemics of those speaking to the media. However, Eisenhower waited a full five days after the launch of Sputnik 1 to respond to the media and he was taken aback by their hostility at his news conference on October 9. His belief was that the early ill-informed press responses such as that of the *Chicago Daily Tribune* on October 6 would dissipate once he provided the calm reassurance that had previously been readily accepted by the Washington press corps—and therefore, by their readers.[56] Indeed, Ike had been used to an easy ride from the press throughout his presidency. He was still revered as the great war leader and the man who had brought an end to the war in Korea within six months of assuming office. He had presided over a period of strong economic growth and his approval ratings remained high. But on October 9, he faced journalists prepared to question the administration's view that Sputnik was just "a silly bauble," prompted by political critics of the administration such as Democrat Senators Stuart Symington, previously secretary to the Air Force under Truman, and Senate Majority Leader Johnson who had already been publically critical of the administration's lack of response.[57] The United Press's syndicated story on Sputnik dated October 5, and published across the country the following day had been headlined: "President not upset over Russian satellite but aides show concern."[58] The story quoted Symington and reported:

> Senator Stuart Symington (D-Mo) one of the sharpest critics of this country's missile program, said that Russia's launching of an earth satellite "is but more proof of growing Communist superiority in the all-important missile field."

> Symington sent a telegram to Senator Richard Russell (D-Ga) calling for an investigation of the US missile program by the Senate Armed Service Committee which Russell heads.
>
> He said that the investigation would give Americans the truth about the US missile program, "something they have not been getting."
>
> Symington said that "while we continue to learn" of Russia's missiles, "this government continues to cut back and slow down its own missile program."

Symington, Like Johnson, was positioning himself as a potential Democratic Party contender for the 1960 presidential election. But he lacked the senatorial power of Johnson who began to appear in the news coverage of Sputnik himself only on October 8, following Russell's agreement to launch a preliminary congressional inquiry, as per Symington's request.[59] Johnson was certainly not the first Democrat to use Sputnik as a potential platform to raise his profile higher, but the fact that his subcommittee of the Armed Services Committee was interested in reviewing the US missile program definitely gave the journalists a new line of questioning about the congressional challenge for the president at his October 9 news conference.

Whitman's diary entry for October 9 commented on Eisenhower's early-morning preparation for the news conference.[60] Meanwhile a series of drafts of statements to be used in that the press conference make it abundantly clear that while Eisenhower may have been playing down the impact of Sputnik, he was well aware of the probability that the reaction to Sputnik could escalate into wild speculation at the hands of journalists sensing a massive story, as well as from those within the military-industrial complex who saw the Russian satellite success as the spur they needed to press for higher spending on defense.[61] Whitman noted that the president was in his office "much too early—7.37am—and immediately started dictating about outer space." She commented on the swearing-in ceremony for Neil McElroy who had succeeded Charles Wilson as secretary of defense (one would have expected this to be the most important event of the day), but also noted that immediately after the ceremony, "certain of high defense officials were in President's office for about 15 minutes to discuss the Russian satellite, more as a preparation for the President for the press conference than anything else, I believe."

The tone of the news conference is quite different from others across the Eisenhower presidency. While the transcript clearly shows the 1950s politeness and deference the press was used to operating by, there was an edge of hostility, and there were some sharp questions that Eisenhower did not deal with particularly effectively.[62] United Press International reporter Merriman Smith, noting the satellite launch and the Soviet claim to have successfully launched an ICBM—both ahead of the United States—challenged the president as to what he was going to do about it. Robert Clark of Associated Press echoed his colleague, asking if the Russians were now ahead of the Americans. "Miss May" Craig, Washington correspondent for the *Portland Marine* asked Eisenhower if the satellite gave the Russians the ability to launch missiles from platforms in space, while NBC's Hazel Markel asked: "Are you saying at this time, with the Russian satellite whirling about the world, you are not more concerned or overly concerned about our nation's security?"[63]

Eisenhower's response was calm—as had been his responses to the previous questions. He denied the link between the satellite and ICBMs; he downplayed the Soviet satellite advantage, though he acknowledged it as a psychological success. He had attempted to allay Miss Craig's fears by stating that the satellite was most certainly not a nuclear missile platform, and sought to allay those of the whole nation. He said: "I see nothing at this moment, at this stage of development that is significant in that development as far as security is concerned." Indeed, he felt that the Soviets had done little more than "put one small ball in the air."[64]

In his news conference, Eisenhower was clearly aiming at calm reassurance. Yet his responses seemed a little flippant, one would suggest because the assembled journalists had not been immediately prepared to accept his statement on Sputnik and had continued to press him on the issue. He was unshaken in policy terms, but had not yet gained the high ground with the media. He was not helped that day by a misalignment that ensured his message, and that of new Secretary of Defense McElroy were out of step. Whitman recorded that McElroy "held a press conference today [thus competing in effect with the President's press conference] and while the President said that the Russian success with the satellite would not cause us to speed up our missile program, Mr. McElroy said in effect that the program would [Whitman's underlining] be speeded up."[65] What is interesting is how McElroy, intentionally or not, immediately disregarded what others would have taken as an order from the president in that meeting with advisers which McElroy attended. Goodpaster wrote:

> [Eisenhower] recalled that there had been a definite and intentional separation of the military and scientific lines of effort with the satellite proceeding as a scientific project.
>
> When military people began to talk about this matter . . . they tend to make this matter look like a "race" which is exactly the wrong impression.
>
> He ended by saying he wanted to enlist the efforts of the whole group in behalf of "no comment" on this development. There was an indication from the group of understanding and support of what the President wanted.[66]

McElroy was brand new to the Pentagon having, like many figures in the Eisenhower cabinet, joined from industry. He had no background of military service, indeed he had most recently been president of Procter &

Gamble. His comment on speeding up missile production was ultimately proven to be correct, but at this time when Eisenhower, as chief executive, wanted all his direct report cabinet members speaking as one, McElroy's responses were distinctly off-message.

Evidence such as the *New York Times* case study from earlier in this chapter point to a change in tone by the newspapers after Eisenhower's October 9 news conference. Sputnik was no longer "new news." And while it remained a useful subject for the news agencies to syndicate in their circulation battles, the story began to move from the news pages further back through the papers. Taking one "heartland" example, the *Kansas City Star*, from Kansas City, Missouri, led with "World Title to the Braves" on page one, but there was a down-page article detailing how readers could spot "Sputnik's burnt-out rocket" as it orbited over the United States.[67] In six days, the newspaper perception of Sputnik had changed from a threat to a curiosity. Eisenhower featured on page three, but only in welcoming the prime minister of Ghana to the White House as his breakfast guest. By the following day, the paper had localized the Sputnik story with two page one pieces. The first "'Beep' a tocsin for education" reported on the chancellor of the University of Kansas' pleas for a crash program to train more scientists. It was a popular call at the time, but a somewhat opportunistic plea from an educator. The second story "Set for Sputnik party" reported on how a local retailer had recorded the Sputnik beep onto vinyl records and was selling them for $2.95.[68] By Sunday, October 13, Sputnik had disappeared from the news pages, which were led by Queen Elizabeth II's arrival in the United States, and the birth of a baby hippo at Swope Park Zoo.[69]

THE RESPONSE OF TV AND NEWS MAGAZINES

Eisenhower faced a second wave of potentially difficult media coverage on Sputnik: that of the news weeklies and television news talk shows which had to wait until their regular publication and scheduling slot to comment on the Soviet satellite. As a prime example of the latter, on October 12, retired Air Force General George Kenney talked to Mike Wallace on the top-rated ABC show *Interview*.[70] Kenney had been MacArthur's senior Air Force commander as commanding general, Allied air forces, Southwest Pacific in World War II, and came to the TV studio as a war hero and renowned defense analyst.[71] Yet his interpretation of the significance of Sputnik was riddled with misunderstanding and misinterpretation, and merely repeated—with little added depth—the newspaper headlines from the previous week. His core argument was that the successful launch of Sputnik 1 proved that the USSR had developed the rocket technology necessary to propel an ICBM into US air space, posing a serious threat to the security of the country. Kenney argued that it demonstrated that America had been too complacent and apathetic about the Soviet ability to develop weapons and produce them in quantity. He claimed that such apathy had given the Russians a lead in the nuclear race, and the day the Soviet political and military staff decided they could win a nuclear war, they would "pull the trigger." Thus war was now demonstrably closer. He added that the United States had neglected the ICBM problem and stated that the United States was now behind the Soviet Union in nuclear weapon development because the American public, whose opinion, according to Kenney, steered defense policy, had not taken the threat seriously enough.

Kenney was quick to place the blame on the American people rather than Eisenhower for the perceived defense shortcomings, noting that they elected all those responsible for making defense and foreign policy decisions, and those decision makers were driven by the public appetite for: "a balanced budget, lower taxes, bigger Social Security benefits, more pensions and better roads." He was actually quite measured in his argument, but was consistently prompted by Wallace, the presenter, who set up the interview by describing the post-Sputnik period as "A struggle for control of outer space." Certainly, by the end of the program, the viewing public would not have been reassured by the key message emerging from this prime-time news leader: World War III is now closer.

The news weeklies initially offered a similarly portentous tone. This was aptly summed up by the editorial in *Time* on October 14, entitled *Red Moon over the US*, which stated that Sputnik:

> opened a bright new chapter in mankind's conquest of the natural environment and a grim new chapter in the Cold War . . . Russia's victory in the satellite race proved that the US had not tried hard enough.[72]

These were leaden-sounding words, but what did they actually mean? At first reading, they seem critical of the president—for "the US had not tried hard enough," it is easy to read that Eisenhower had not tried hard enough. But was the "grim new chapter in the Cold War" a Russian satellite that did nothing more than go beep, or could it have been the demands of the retired generals and grandstanding politicians for the United States to enter a phase of crash spending on nuclear armaments? While scholars have tended to read a single meaning into the post-Sputnik media coverage, there have been few attempts to chart the changing tone of the coverage.

It is clear that over the fortnight following the news the Sputnik 1 was in orbit, those pushing a particular anti-administration agenda used the media to further their cause. Often the messages were relatively subtly planted, but pushed the message hard that the United States needed to ramp up its defense spending. For instance, the *Kansas City Star* did return to the Sputnik issue to run a long editorial piece on October 20 discussing the view headlined: "Sputnik the result of Soviet science system."[73] The piece noted that the USSR spent 40 percent more on atomic research than the United States, and urged the government not just to meet but to outstrip this sum. On the same weekend, a widely-circulated Associated Press piece forecast additional defense spending, no tax cuts and stiffer credit controls as a result of Sputnik.[74] Yet, both the case for increasing basic US research and for ramping up defense spending were part of Eisenhower's existing policy stemming from the TCP recommendations in 1955. They were neither created nor shaped by Sputnik. Indeed, the so-called Sputnik crisis seemed less of a crisis day by day by day.

While Sputnik had dominated the presidential press conference on October 9, it was relegated to item four on the agenda and gathered little interest at the next presidential news conference on October 30.[75] It is notable too that there was no clamor from the White House beat reporters for an additional press conference in the intervening weeks. While Sputnik was the main topic of discussion at the presidential news conference on October 9, it was not the only one. The exchange with the press also featured two questions on the Little Rock school desegregation issue.[76] By October 30, the mood had changed considerably. The news conference was wide-ranging, opening with a comment on the bombing of the Israeli cabinet before taking in the upcoming NATO annual meeting in Paris, bilateral conversations with Macmillan, segregation in the South, the rise in the cost of living, prospects for the stock market and Eisenhower's views on who should be the next governor of New Jersey. In short, it was back to business as usual.[77]

OPPORTUNISTIC ATTACKS

Immediately following the announcement of Sputnik's orbit in the United States, there was a high degree of bandwaggoning from those in the military and political establishments who saw outer space as an opportunity to advance their own agendas. US Senate majority leader, Lyndon Johnson, wrote what has subsequently been described as a diary entry, but one that was made public first in his Presidential memoir *The Vantage Point*: "Now, somehow, in some new way the sky seemed almost alien. I also remember the profound shock of realizing that it might be possible for another nation to achieve technological superiority over this great country of ours."[78] General James Gavin, a member of the US ICBM development team described Sputnik as "a technological Pearl Harbor," though of course, it provided a great lever to swiftly place the US ICBM program at the top of the president's agenda. Symington was still using the same line a month later as he sought to establish his own space credentials in preparation for a possible challenge for the White House in

1960.[79] Indeed, the perception of the Soviet success both in the Pentagon and on Capitol Hill was not a scientific triumph but a militaristic threat—which played directly into the hands of the "hawks" of the military-industrial complex.

The US Navy, too, was not slow in issuing statements to the media that mixed scare tactics with shameless promotion of its own strength. However, by October 9, even as Eisenhower prepared to address the White House press corps, the tone was changing with rather more perceptive comment and questioning emerging in the US media—even in the rural heartland of the Midwest. On October 9, the Opinion, in the *Salina Journal,* the closest daily newspaper to Eisenhower's boyhood home in Abilene, Kansas, stated:

> At this writing, it appears the Russians and the U. S. Navy have won the Sputnik contest. The Russians have floated their basketball in space, thereby deflating our ego. . . . The U. S. Navy won because its bright public relations experts rushed into print with the assertion that the Russians also have intermediate range missiles capable of placing nuclear warheads on every Allied base in western Europe. The Navy contends that while this makes our overseas Air Force bases worthless, it enhances the value of Navy aircraft carriers which move about on the deep too fast to become missile targets. . . . That also provokes sober thought. Were it not for the costly, senseless, bitter rivalry between the Navy, the Air Force and the Army, it is quite likely we would be ahead of the Russians, not behind them.[80]

PYRRHIC VICTORY

The success of the Sputnik 1 mission won a "race" of propaganda and prestige for the USSR, but the victory was Pyrrhic and even by the time Sputnik 2 became the second man-made object in space, the Soviets' success was already beginning to tarnish. The shock of the new was a once-only effect, and the American media's response to Sputnik 2 was less fearful and less polemical. However, it would appear that the traditional scholarly narrative of the effect of the first Sputnik launch on the American public does not quite reflect the historical reality of the reaction. On April 14, 1958, Oliver Gale, special assistant to the defense secretary wrote to McElroy with some findings gleaned from Claude Robinson's public opinion index.[81] Robinson was George Gallup's partner in their pioneering market research business, while Gale had set up Procter & Gamble's first public relations department and had moved to the Pentagon when McElroy, the Procter & Gamble president, moved to the DoD. Gale noted Robinson's sample survey of one thousand adults in the general public and 117 newspaper editors made immediately after the launch of Sputnik 1. Under the heading "Impact of Sputnik," the memorandum noted:

> The news of Sputnik's launching was known to 95 percent of the public, but 40 percent noted the news and dismissed it without serious thought as to what it might mean to them and their country.
>
> After Sputnik, 80 percent thought we were "at least even" with the Soviets or "would catch up before long"; editors had a more alarmed point of view.[82]

This simply does not reflect a sense of panic. Gale noted the difference between the public, where the response was somewhat sanguine, and the "more alarmed" view of editors. Yet this differentiation is too often overlooked in the narrative. Roger Launius's summation of the launch and Eisenhower's reaction reflects an orthodox approach:

> During the furor that followed *Sputnik 1* and *Sputnik 2,* many people accused the Eisenhower administration of letting the Soviet Union best the United States. The Sputnik crisis reinforced for many people the popular conception that Eisenhower was a smiling incompetent; it was another instance of a "do-nothing," golf-playing President mismanaging events.[83]

This is not true.

EISENHOWER'S MEDIA OFFENSIVE

Sputnik 2 appeared to be another propaganda coup for Khrushchev, putting the dog Laika into orbit just a month after Sputnik 1's launch. Behind the scenes, Eisenhower had been maneuvering since the October 9 news conference to educate the public in how to deal with the new reality of artificial satellites. His position remained unchanged from the first Sputnik: he saw no indication from Sputnik 2 that the Soviets actually had a workable, accurate and reliable ICBM, and urged both Nixon and Dulles to stress these points. Indeed, Nixon had used this argument in a public speech in San Francisco on October 15, and Dulles in a meeting with the press a day later.[84] But the information released by the Soviets appeared to challenge the administration's thinking. Sputnik 2 carried a 1,121 lb payload, which underscored the strength of the Soviet rockets. The fact its payload was a dog implied the Soviets were focused on manned space flight.

This time, rather than play down the significance, Eisenhower chose to go on the offensive and address the nation via television on November 7. He addressed the fear in people's minds about national security. In seeking to soothe the nation, he said: "We are well ahead of the Soviets in the nuclear field both in quantity and quality. We intend to stay ahead."[85] Here Eisenhower, advised by Hagerty, regained the media initiative by addressing the US public directly rather than allowing journalists to editorialize around his words, and did so quickly before rumor could fill any fact-free vacuum. Such a response played well with the news magazines, but they did not immediately side with the president. Their longer publication cycles allowed some to feature both the Soviet propagandizing of Sputnik 2 and the administration's response.

For example, *Time* was the most hyperbolic in its response, especially when it noted: "The Soviet rocket generated a total thrust more than enough to power an atomic bomb to the moon, more than enough to power a missile around the earth. . . . In such an apocalyptic week, communism's new coalition of dazzling technology and cut-throat politics represented an epochal threat to the free world."[86] *The New Republic* was rather more circumspect and stated:

> The tendency to over-react to sputniks by concentrating on missiles to the jeopardy of all else (or, to put it more precisely, to react in the wrong ways) can be held in check by bearing in mind that it is a long way from the successful test firing of an intercontinental ballistic missile to an operational system. That was so the day after Sputnik 1 went up, and it still is.[87]

Internationally, however, even journals such as *Spaceflight*, the journal of the British Interplanetary Society and thus untouched by the political machinations stirred up by Sputnik, added warnings to the president on the apparent power of the Soviet rockets. "It is . . . logical to assume that at least in the early development stage (the rocket) was a military project, having as its goal the delivery of an H-bomb warhead over perhaps 5,000 miles."[88] But the purely negative sentiments of the media were short-lived. This was to a large extent down to the more active and considered stance that Eisenhower took in the wake of the second Soviet satellite. His action had ensured the media tide began to turn in his favor.

It was not, however, an immediate 180 degree turn and for a while, the nation—as exemplified by the press—did not know what to think. The November 18, 1957, issue of *Life* provided a prime example of the almost schizophrenic response of the news magazine editors to the new attack on US confidence.[89] Within this single issue, *Life* praised Eisenhower for his November 7 radio and TV address, and was positive about his appointment of Killian as special scientific adviser. It interviewed influential figures on the president's speech, eliciting mixed reactions: two positive and two negative. Investment Banker Ferdinand Eberhardt felt the president's planned actions post-Sputnik "may prove inadequate to meet the needs of our present critical situation." Unsurprisingly, Eisenhower's former rival for the White House, Adlai Stevenson, was also critical: [Eisenhower's speech had] "too little sense of urgency." But seemingly in the spirit of balance, the magazine published two pro-administration views. Paul Foote, assistant secretary for research at the Department of Defense found the Killian appointment "a good first step to having a Secretary of Science in the Cabinet, a move he favors." The president of defense contractor Boeing, William Allen, echoed Foote's praise for the appointment of Killian stating: "Now is the time to speak of what we need, not what we have."[90]

If the November 7 TV and radio broadcast had begun Eisenhower's fight back, that fight gained momentum in the second of three planned addresses to the American people. Titled "Our Future Security," the speech, which was also carried on radio and television, was made in Oklahoma City on November 13 as Eisenhower joined the Oklahoma state leaders to celebrate the fiftieth anniversary of Oklahoma's statehood.[91] This speech set out clearly Eisenhower's view of security and his undiminished desire for fiscal responsibility—in fact, it is a microcosm of Eisenhower's driving tenets: peace through national security and fiscal austerity. This was not a particularly rousing or glamorous speech, but after the reassurance of the November 7 address, it did much to articulate the Eisenhower vision for security, and thus explain his resistance to the clamor for ever-increased defense and rocket spending. Giving the speech, Eisenhower sounded far more comfortable delivering an address in the "Heartland" than in the earlier address which was made from the Oval Office. The speech was written very much as a joint effort between Eisenhower and former director of the US Information Agency, Arthur Larson, whom Eisenhower had brought into the White House as a special assistant specifically for the purpose of drafting speeches on science and national security policy. It underwent much iteration with Eisenhower's notes and changes visible across each of the drafts as the speech developed.[92] Eisenhower repeatedly cut out the metaphors and replaced high-flown sentiment with the kind of plain speaking he was comfortable in delivering. The result was a very Midwestern style of delivery. Eisenhower also instructed Larson to remove any specific references to attacks by his political opponents—he did not need to give them any further exposure.

The main theme of the speech was security. Eisenhower first acknowledged the Soviets' "rigorous educational system and their technological achievements." But he warned that this achievement was "happening under a political philosophy that postpones again and again the promise to each man that he will be allowed to be himself, and to enjoy, according to his own desires, the fruits of his own labor." Having reminded his audience of his anti-communist credentials, he was eager to remind America of its strength: "It is found in the quality of our life, and the vigor of our ideals. It manifests itself in the ever-astonishing capacity of free men for voluntary heroism, sacrifice and accomplishment when the chips are down." Eisenhower went on to provide a lesson to his audience, that the Soviets were quite willing to launch "aggression by violence [but also by] . . . propaganda and subversion, economic penetration and exploitation." He then set out America's defense, stating that it was based on "a strong nuclear retaliatory power [and] . . . co-operation with our allies." The lesson stated the costs of different elements of defense—both those he supported and those demanded by his political opponents and military commanders. In five years, he said, his administration had spent $211 billion on national defense. He acknowledged the calls for "an indiscriminate increase in every kind of military and scientific expenditure." But then Eisenhower delivered a sober and reasonable assessment of actions that would most likely emerge from the planned national security review—planned as an annual event, and therefore not driven by any perceived need to react directly to Sputnik.

Eisenhower picked up on elements of the still-secret Gaither Report on defensive military capabilities, not referring to the report directly, but talking about speeding up the dispersal of SAC to additional bases; quicker SAC response to emergency alarms; improvements to early warning systems and developing "active missile defense against missiles" as well as adding long-range missiles. The point was to drive home that under Eisenhower's command, his appointees were doing much of what their political opponents had accused them of not doing in the wake of Sputnik. For the president, throwing money at the issue did not necessarily hasten the solution, and while some areas, such as ICBM development, required more funding, that would come only by prioritizing according to need. That, for Eisenhower had to be a balance. For every ICBM program, commensurate cuts needed to be made elsewhere—either in defense or in civilian programs. This was a plainly stated message to Congress. Eisenhower would not tolerate deficit budgeting. "Now, in the Federal government's civilian activities, we shall have to make some tough choices."

As the speech developed, Eisenhower drew it away from additional spending on crash military programs. In making his key argument that the United States should be investing more in scientific education and basic research, he noted that "more" did not necessarily mean more federal funding: "Frequently time is a more valuable coin than money. It takes time for a tree to grow, for an idea to become an accomplishment, for a student to become a scientist." He both asked for greater investment in science education *and* the patience of the American people to let that education take root. In terms of US basic research, he noted that: "compared with any other country's [it] is considerably greater in quantity and certainly equal in quality." But he was aware at "the fast rate of increase in the Soviet effort and their obvious determination to concentrate heavily on basic research." Again, the solution was not money. "You cannot say to a research worker, 'Your salary is tripled, get busy now and produce three times as many basic discoveries.'"

This speech reassured the US public that the nation had the defenses in place to withstand any Soviet threat, but also pushed back the arguments of Eisenhower's post-Sputnik detractors by stating that his administration was doing exactly what they had stated he wasn't doing. SAC was being strengthened; missiles were being developed and all this was being achieved in the American spirit of openness and freedom for the individual to succeed. Any dramatic acceleration in any of the programs mentioned would require cuts elsewhere, and the challenge was for Congress to identify and deliver these. It was rather a clever political speech just a year out from midterm elections. This was a speech of calm assurance, a confidant rebuttal from a leader with no need to fear either the Soviets or his own domestic detractors.

Once the November 7 date had been set to counter any hostile reaction to the launch of Sputnik 2, the Oklahoma City speech evolved to be the middle element of a three-part response to Eisenhower's critics, with the final address set for Cleveland, tentatively planned for November 29.[93] This concluding element was never delivered due to Eisenhower's stroke on November 25. However, his policy was clear. With the Oklahoma City speech in particular, Eisenhower delivered a series of points that were all consistent with his national security policy from considerably before the Sputnik 1 mission.

THE DEAD DOG BOUNCE

The final point to consider here is the newspaper reaction to Sputnik 2. The launch of Sputnik 2 is often conflated with Sputnik 1 and there is no separate investigation into the way the newspaper reaction to the second satellite differed from the first. In *Eisenhower in War and Peace,* for instance, Smith covered both in a single paragraph which concluded:

> Sputnik orbited the earth 560 miles up, travelling at a speed of eighteen thousand miles an hour. Shortly afterward came Sputnik II, launched on November 3, six times larger than its predecessor with an orbit even higher.[94]

The public and media reaction was very different though, and reflected increasingly the reassurance Eisenhower's active presence was making, and the growing belief that what he had said in his news conference on October 9 was actually true.

On Monday November 4, 1957, the *Chicago Daily Tribune* led with the banner headline: "Dog Rocket Speeds in Sky."[95] Its approach was consistent with the United States' other major regional daily papers. The article focused on the Soviet success in putting the animal into orbit and the page one lead implied that the Soviets had said the dog would be brought back alive. Yet, while the article carried the additional information furnished by the Soviets this time round, notably the size and weight of the satellite and the details of the dog, initially named "Curly" in the American media, there was equal weight given to the US scientist view that it was doubtful that the dog could be returned. In fact, the piece noted that "Curly" was breathing out carbon dioxide which would ultimately kill it.

The *Tribune* continued its news coverage on page three with several short pieces considering the political impact of this latest launch. Edward Teller, whose hawkish reaction to the initial Sputnik launch had shown him no supporter of Eisenhower's more outwardly-moderate view of US-Soviet relations, was quoted as saying: "Russians have ways of imposing instruction and production that we do not have. It's everybody's job in the US to ensure we do not fall behind in technology." This was a considerable step down from the short-lived hyperbole of the previous month. Further coverage on pages three and five of the paper noted no reaction from the White House and Republicans admitting no surprise about the new Soviet satellite. Partisan Democrats though remained critical: Senator Jackson (D-WA) of the Senate Armed Services Committee was quoted by the *Tribune* saying that the administration "should appoint an all-powerful missiles chief and undertake immediately a 'bold program' for increasing scientific and engineering training."[96]

This was about as polemical as the Sputnik 2 coverage got in the newspapers. Indeed, there was rather a twist in the tale for the Soviets as even on the first day of launch coverage, the Soviet scientific and technological triumph ran into trouble with a very powerful lobby they would simply be unaccustomed to in Moscow: animal lovers. In contrast to the pop science hyperbole that followed the launch of Sputnik 1, the *Chicago Daily Tribune* focused on a very "human interest" aspect of the flight. In a page lead entitled: "ASPCA assails Russian sending of dog out into space." Warren W McSpadden, general manager of the American Society for the Prevention of Cruelty to Animals stated:

> The ASPCA deplores the use of a dog or any other living animal, in these earth-launched experiments, whether by the Russian government, reported in the press as already having taken place, or by any other government. Such use will, the ASPCA believes, result only in the unnecessary sacrifice of the animal, or in great pain or suffering should it survive. Moreover, sending a dog to outer space cannot possibly advance human health and welfare.[97]

By the following day, Sputnik was relegated to four paragraphs down-page on page one with the start of the American fight back. The story quoted Dan Kimball, head of US Aeronautics and Aerospace company Aerojet saying America had a rocket capable of putting a half ton satellite in space.[98] Yet this piece still had to take second place to Chicago Mayor Daly's new housing plans. Without a local angle, space news lost out to a local story. A day later, the federal midterm and gubernatorial elections had overtaken the Sputnik story as the paper's main news with the headline reflecting Democrat, Governor Robert B Meyner's overwhelming re-election in New York—a major domestic political blow to Eisenhower and Nixon.[99] The president's decision to address the country on national security shared the front page—but there was clearly no sense of panic about Sputnik 2. By November 7, the paper was speculating that Eisenhower was considering launching a larger rocket than the Russians, despite no word whatsoever emanating from the White House, while its more outlandish extrapolation stated that the Soviets would be on the moon by 1967.[100]

On November 8, the key Sputnik related stories in the Democratic *Atlanta Constitution* noted that the president's TV address had been "aimed at reassurance" and that he had used it to announce James Killian of MIT as "Science Czar." Just four days after the launch, this news was carried only on page eighteen and given equal treatment to the ongoing human (animal) interest story, "Space Dog Feared Doomed."[101] However, there was simply too much domestic US news for the Sputnik 2 story to gain as much propaganda value for the Soviets as the Sputnik 1 launch a month before. The element of shock and surprise had gone. This launch was expected, and events from the gubernatorial and senatorial elections; follow-up to the new British Prime Minister Harold Macmillan's visit, and the search for wreckage of a downed San Francisco-to-Honolulu bound airliner with forty-four aboard all vied for a place in the news pages both of the *Atlanta Constitution*, and newspapers across the country.

MISSILE SUCCESS

Part of Eisenhower's sangfroid in the face of the first Soviet satellite launch was no doubt due to the increasing success of the Atlas and Titan ICBM and Thor IRBM testing by the Air Force as well as the Jupiter competitor missile developed by the Army. The USAF missile development program, under Schriever, was moving ahead rapidly. As often happened with new missiles, the first test flights of both Thor and Atlas were failures. The first Thor flight on January 25, 1957, ended just eighteen inches above the launch pad. Repeated failures followed, but by September 20, Thor 105 deposited a dummy warhead 1,495 miles from its launch site and on October 11, just a week after Sputnik 1, Thor 106 delivered its warhead along a perfect 1,725 mile course before it splashed into the Caribbean near Venezuela.[102] Atlas faced similar teething problems. On June 11, 1957, the first Atlas "A" test vehicle was launched from Cape Canaveral. Standing seventy-five feet and one inch tall, the missile drew thousands of spectators to the Cape to see its launch.[103] The "flight" lasted just twenty-four seconds, covering around 1.6 miles before a range safety officer ordered its destruction due to a fault. The launch was thus only a partial success as was the follow-up test in September. However, in both cases, the faults that had led to failure were pinpointed quickly.

The June failure was down to a failure in the turbo pump, while a failed liquid oxygen regulator caused the second test to be aborted. Both could be fixed quickly as Schriever told the Joint Chiefs of Staff who, in turn, informed the president. The remedies were proven in the faultless flight of the third Atlas test vehicle on December 17, 1957.

The Army's Jupiter Medium Range Ballistic Missile (MRBM) had already undergone two successful tests before Sputnik, while the Titan multi-stage ICBM was slated for first testing at the start of 1959, with the solid rocket fueled Minuteman and sea-borne launched Polaris ICBM both reaching the end of their design stages. Even on the civilianized side of the rocket program, the Navy's new Vanguard rocket, designed to launch the United States' IGY satellite had enjoyed testing success. On October 23, 1957, the US Naval Research Laboratory (NRL) successfully tested a three-stage Vanguard rocket. There were two test launches before the October 23 event. The first test vehicle, TV-0, was launched on December 8th, 1956. It tested the rocket's telemetry systems. Then, on May 1, 1957, the TV-1 tested the separation and subsequent second-stage ignition capabilities of the two-stage rocket design. While there were a number of aborted launches, all three of these tests were successful, further reinforcing Eisenhower's confidence that the United States' military and civilian rocket programs would ultimately be successful. The success of the missile program testing was a key factor in Eisenhower's measured response to Sputnik and it was undoubtedly the new phase in this military program that was most important for the president as he sought to disentangle the military and non-military elements of the emerging space program. The ongoing successes of the military missile development program underpinned Eisenhower's calm in the wake of the Sputnik launch and his refusal to be drawn into either speculation or any kind of hyperbole at the October 9 news conference where the tone of questioning had drawn a dividing line between the media and the administration. The White House remained calm but with no wish to share national security secrets, while the press and broadcast media sold stories on the basis of speculation, ill-informed comments from those outside the administration's inner circle, and tactical attacks from those looking to use space for their own political or economic advantage.

RECOVERY: GAITHER AND EXECUTIVE PRIVILEGE

In December, when the initial interest in Sputnik had faded, the recommendations of the Gaither Report were leaked. This was a highly political action that could have crippled Eisenhower's presidency. Yet his actions in dealing with it reflect the ascendancy Eisenhower was regaining over those who criticized his reaction to the Soviet success in space.

The term "Executive Privilege" was coined in the Eisenhower years, not by the president himself, but according to Arthur Schlesinger, by his attorney general, William Rogers, who had succeeded Herb Brownell in the post.[104] The chief issue he dealt with in his term as attorney general was segregated schooling. Outer space was most certainly not a major concern, but he did advise the president on dealing with the Gaither Report in the face of significant congressional pressure to publish a document that potentially could add considerable heat, though precious little additional light, as the president sought to impose his will in the immediate post-Sputnik era. The TCP findings in 1955 had catalyzed action on the US missile program, on the U-2 spy plane, and on CORONA. The panel system had worked well, and in 1957, it made sense for Eisenhower to authorize a similar confidential study into the United States' passive defense systems. The study was prompted by a report from the Civil Defense Administration recommending a $40 billion appropriation for bomb shelters.[105] Rowan Gaither, chairman of the board of the Ford Foundation was, at Killian's recommendation, appointed to chair the civil defense study. But Gaither became ill almost as soon as the task force commenced its work. William Foster, a former secretary of defense, took his place as co-director, alongside Robert Sprague, president of the defense contractor, the Sprague Electric Company. Eisenhower had specifically instructed the task force to focus on passive defense. But by September, there was strong evidence of "mission creep." While one might have expected this to be initiated by Foster, after his years in DoD, the zealot for widening the mandate to include reviewing the nation's active nuclear deterrent was Sprague. He was rather in thrall to RAND staff report R-290, compiled by Albert Wohlstetter in September 1956 and it was Wohlstetter who influenced Sprague to widen the scope of the study.[106] This claimed that just 150 Soviet ICBMs could wipe out the entire SAC B-52 fleet in a surprise attack.[107] While there were many reasonable elements to Wohlstetter's report, such as the need to disperse the US nuclear bomber force, it was rather apocalyptic in its sentiment. However, the Gaither task force made the most polemic elements the center of their response to the administration. This response gained a rather more hawkish edge in the final drafting which fell to Paul Nitze. Given his background, it is unsurprising that Nitze's draft was unequivocal in its demand for another massive investment in defense spending.[108]

The Gaither committee presented its findings to the president on November 4, a day after Radio Moscow announced the successful orbit of Sputnik 2. The committee reconvened to address the NSC three days later. The report had been through many hands in the Pentagon prior to the NSC meeting and appeared to provide significant ammunition for those in the defense lobby looking to see significantly increased spending both on SAC's bomber force and more starkly, a vast increase in both the number of ICBMs and IRBMs—and the rate at which they should be produced. Titled *Deterrence and Survival in the Nuclear Age* the report devoted relatively little space to the committee's original remit of reflecting on the state of US passive defense.[109] It was startling in its demand to offset the perceived threat of Soviet ICBMs by building an ever-larger deterrent force. Perhaps the most telling section is how this would be funded. If ever a report supposedly prepared for the benefit of the president misjudged both the needs and the nature of the president, it was Gaither. On page four of the report, the committee laid out its estimate of the Soviet threat:

> The singleness of purpose with which they have pressed their military-centered industrial progress has led to spectacular success. They have developed a spectrum of A- and H-bombs and produced fissionable material sufficient for at least 1500 nuclear weapons. They created, from scratch, a long-range air force with 1500 B29-type bombers; they then substantially reequipped it with jet aircraft, while developing a short-range air force with 3000 jet bombers. In the field of ballistic missiles they have weapons of 700 n.m. [nautical miles] range; in production for at least a year, and probably surpassed us in ICBM development. They have developed air to surface and probably submarine-launched cruise missiles; built 250-300 new long-range submarines, and partially modernized 200 others. They have created an air defense system composed of 1500 all-weather and 8500 day jet fighters, equipped at least 60 sites, each with 60 launchers, for a total of 3600 launching pads for surface-to-air missiles provided with a sophisticated and original guidance system and a ground environment of 4000 radars. At the same time they have maintained and largely reequipped their army of 175 line divisions while furnishing large quantities of military equipment to their satellites and Red China.[110]

The figures, lifted largely without question from the RAND study, were without any significant substantiation. The panel had not, of course, had access to Eisenhower's reports on the U-2 over-flights of Soviet territory or other CIA material. These told a very different story. There was indeed a weapons and missile gap between the United States and USSR—but it was hugely stacked in America's favor. If they had rather overemphasized the scale of the Soviet threat, the committee demonstrated just how out of touch it was with the president's thinking when it came to fund the necessary increase in defense spending to deal with the perceived shortfall in US preparedness.

> The added defense measures to which the Panel has assigned relative values will probably involve expenditures in excess of the current $38 billion defense budget.
>
> The measures of highest value to strengthen our deterrent and offensive capabilities are estimated to cost over the next five years (1959–1963) a total of $19 billions.

> Additional measures of somewhat lower than highest value, for the protection of the civilian population include a strengthening of active defenses, a fallout shelter program and the development of a defense system to protect cities from missile attack. The estimated cost of these items total a cost of $25 billions over the next five years. . . .
>
> The American people have always been ready to shoulder heavy costs of their defense when convinced of their necessity.[111]

Eisenhower was not convinced. He recognized the good sense in dispersing SAC's bomber force and was prepared to go along with the Air Force's drive to speed up the ICBM/IRBM program which had been lagging under Wilson's tenure at DoD (not least because Eisenhower had insisted on reducing defense spending). However, he had no intention of vastly increasing offensive military spending to balance out a threat he knew to be wildly exaggerated. In fact, his response to Gaither is a fair reflection of the Eisenhower presidency. Ever since the Korean armistice, he had concentrated on finding means to reduce defense spending while building just enough nuclear threat to meet the counter-threat posed by the Soviets and latterly the Chinese. Increasingly that expenditure had moved from large ground forces, an over-sized navy, and the massive retaliatory threat of the SAC to a focus on a smaller, nimbler force for policing actions, and a missile-based nuclear deterrent. Now the rather misinformed rhetoric of the Nitze-crafted Gaither Report threatened the twin tenets of Eisenhower's strategic view: fiscal conservatism and an innate desire to maintain peace with the Soviets.

Eisenhower's response to his oral briefing on November 4 and the extremely well-attended NSC briefing three days later was a classic of presidential style. S. Everett Gleason, who minuted the meeting caught Eisenhower's tone superbly. He was not going to panic nor bend to the Nitze-led argument. He noted that he had advocated an increase in the defense budget from $38 billion to $39.5 billion, but that this had been rejected by Congress. He then questioned whether the evidence presented was done so in order to protect the United States against a real Soviet threat, or merely to enable the maximum number of people in the defense and related industries to make as much money as possible out of the situation:

> The President added that in the light of what had been presented at the Council meeting today it was essential that we neither become panicked nor allow ourselves to be complacent. It was necessary urgently to make an economic, psychological, and political survey of what could and should be done. In this context, perhaps the advent of Sputnik had been helpful. The President added that we certainly did not wish to appear frightened. . . . The President believed that we could correct this situation. The problem was whether we could correct the tendency of every American to try to get the maximum for himself out of the operation of our free economy.[112]

In discussing the impact of the proposed massive increase in both passive and offensive defense spending on foreign relations, John Foster Dulles made it abundantly clear that the president was not going to act on the major recommendations of the report, but that the committee members were going to have to come to that realization rather than get a straight rebuttal from the president. One point was key: the struggle with the Soviet system was not only a military contest.

> Secretary Dulles then returned to the point that our struggle with the Soviet Union and international communism was not just a military struggle. Up to now it has been primarily a cold war. Accordingly, there was great danger that we should so focus our eyes on the military aspects of the struggle that we lose the cold war which is actually being waged, forgetting that an actual military conflict may never be waged. The Soviet Union could make enormous gains in the economic struggle between us if the United States devotes so much of its resources to military measures and shelter programs that no resources remain for waging and winning the cold war.[113]

As a further step towards burying the unacceptable, Eisenhower asked the various agencies present at the meeting to study the report and report back on any of the recommendations they felt worthy of further consideration. It was a rather charming way of letting the committee know that nothing substantial would be implemented. And indeed, he concluded by stating that "he could not thank the Security Resources Panel and its Advisory Group enough for the marvelous piece of work they had done."[114] In normal circumstances, that would have been that. The task force had been assembled at the president's request. They had made their report and the president had listened. As far as Eisenhower was concerned, that was the end of the Gaither process. However, Eisenhower's opponents, Democrats such as Nitze for instance, had too much information that could be damaging to the current administration to let such a potentially incendiary report lie.

A QUESTION OF TRUST

One of the enduring strengths of Eisenhower's relationship with his chosen advisers was his utter faith in their integrity. It was Eisenhower's view that those close to the president should not leak information unless he directly told them to. In practice, such discretion and devotion to the administration's cause had ensured the efficient implementation of Killian's TCP report enabling such programs as CORONA and the U-2 to still be very secret even across the Washington political village. But someone close to Gaither, almost certainly Robert Sprague, quite possibly in conjunction with Nitze, was talking to the press. On December 20, just two weeks after the Vanguard test failure, and while Eisenhower was still recovering from his stroke, Chalmers Roberts wrote an in-depth piece summarizing the report's recommendations in the *Washington Post.*[115] Publication in the *Post* ensured the Gaither summary would be read both nationally and internationally. Democrat Senator Joseph Clark of Pennsylvania then raised the temperature a little higher by inserting Roberts' story in the *Congressional Record.* Particularly through his subcommittee on preparedness, Lyndon Johnson pressed the administration to make at the very least a sanitized version of the Gaither Report (and indeed the previous Killian TCP report) available for scrutiny by Congress. It is questionable whether this would have had any benefit to the nation. However, in pressing for publication, Johnson was making an overt political maneuver both to emphasize his congressional leadership (and potential for future executive leadership), and, more so, to embarrass his Republican opponents. But Eisenhower was adamant that he had no intention of publishing either report. Indeed, in his memoirs, Killian records that:

> The constructive impact of the study was greatly diminished when it was leaked to the press. In the administration, the leak discredited the report, and it doubtless did provide ammunition for opposition politicians who wanted to attack the administration. All of this made it difficult for the President to follow through on those parts of the report of which he approved.[116]

It is interesting that Killian raised these points. Had the report not been leaked, the administration could have claimed credit for any steps it took following the report's receipt and digestion. But the leak provided ammunition for Eisenhower's opponents. They could claim support for *all* Gaither's recommendations and state that the decision of the administration to disregard them was a failure of government. Making the Gaither recommendations public took them away from the president and made the report seem all the more critical of the administration. Eisenhower was stubborn, and though the concept of executive privilege was new, he was firm in his belief that it was the right of a president who commissioned a report to keep its findings to as small a group of insiders as possible and act on those findings as he saw fit. It became apparent that some of those involved in compiling the report were lobbying for its recommendations to be implemented. That was not their role. At the NSC's 350th meeting on January 6, 1958, Eisenhower made his position abundantly clear.

> The President commented that he believed that before we got done with this Gaither thing we would find ourselves obliged to do things which we normally would never think of doing [releasing a classified report to the President prepared confidentially by a board of consultants appointed by the President]. Mr. Cutler expressed his very deep opposition to making any concessions to the demand for versions of

> the Gaither report, and said that what the Congressmen and Senators were most interested in were the timetables in the Gaither report. The President replied in exasperation that he was sick to death of timetables; he had had experience with them for years, and they never proved anything useful. Mr. Cutler repeated his view that even the issuance of a sanitized version would have catastrophic results.[117]

CONCLUSION

For over fifty years, Eisenhower's reaction to Sputnik has too often been built on the assumption that the president was caught cold by the launch of the Soviet satellites and that his reaction was both slow and forced. More recent revisionist interpretations, starting with Divine and continued by Mieczkowski have moved towards detailing Eisenhower's response as both pragmatic and understandable within the limitations of his fiscal austerity and narrow focus on national security. But even in these revisionist observations, Eisenhower, in some respect, fails. For Mieczkowski, one of the key failures was "failing to heed the warnings of a first satellite's prestige."[118] This made his job tougher and opened him to attack from the media and Congress alike. His failure was one of imagination: the ability to link satellite success to enhancing the image of the nation. Mieczkowski also noted that Eisenhower's rigid fiscal conservatism, while eventually enabling him to leave office with the lowest inflation rate among all out-going presidents, absolutely ensured he neither would nor could fund a space race for national prestige.[119] Perhaps this is less a case of failure, than of a hindsight view of missed opportunity. Yet what is striking about even the most recent accounts of Eisenhower's actions on space is the unquestioned assumption that they were a response to Sputnik. The evidence does not support this conclusion. Through the intelligence gathered by the CIA with its U-2 over-flights, Eisenhower knew that Sputnik posed no threat to national security. Through the work of the TCP, the strength of Strategic Air Command and the steady progress on the development of IRBMs and ICBMs, he knew that he had the elements in place to deliver his threat of massive retaliation which provided the fulcrum of the New Look defense policy. As he said himself, Sputnik did not change that "by one iota."[120] His long-term strategy was not even changed by the Gaither panel recommendations. He also knew that as the end of the International Geophysical Year—and its planned climax for the United States with the launch of its own satellite—approached, he needed to take steps to enable the next phase of US space policy to proceed along the parallel tracks of secret military projects and public scientific ones. Consequently, his action was a continuation of this path and the need to react in any meaningful way to the obvious Soviet propaganda play of the Sputnik launches simply was not a factor in his thinking.

It is true that after the first Sputnik launch this rather narrow national security-based thinking did hinder him in his response to the American people and the wider world through the media. But he did seize back the initiative, to the point that Sputnik 2 swiftly became a non-event. What is most compelling is that he won back the initiative by enacting his own preexisting policies, *not* by following the lead of others. What furor there had been in the wake of the Soviet launches had been largely media-manufactured and sustained by those with an agenda, whether that was purely political, or related to defense spending. It is likely that Eisenhower could have headed off that furor sooner had his initial response been less tardy. Staying in Gettysburg playing golf on the weekend following the first Sputnik launch was a poor decision prompted by poor advice from an expert employed to brief him both on the media's needs and their likely reaction to events. Waiting three further days to provide any meaningful response to the growing concern, particularly among the media, was not a show of strength but a lapse in understanding of the impact not of the particular threat the satellite posed, but of the very fact that the Soviets had launched it and launched it first.

But it is wrong to dwell on that early misstep. From October 9 onwards, Eisenhower worked pragmatically to allay the media's fears and those of the wider public. What was crucial was that he was not diverted from his own path on both missile development and in fulfilling US expectations of the IGY. In reality, Sputnik was little more than a minor bump in the road. The political fight against the president's moderation, particularly in defense spending, continued after the initial flurry of Sputnik-inspired publicity, and Eisenhower's poor health at the end of 1957 left a void in the administration's response to criticism. But Ike's stoic determination in following his chosen path for national security, and what that meant for the incipient US space program, ensured the administration's wishes became legislation just nine months after the first Sputnik launch.

NOTES

1. Murray and Bly Cox, *Apollo,* p. 23.
2. Ibid., p. 24.
3. Tindall and Shi, *America,* p. 1214.
4. M Brzezinski, *Red Moon Rising* (London, 2007), p. 173.
5. R Launius, "Kennedy's Space Policy Reconsidered," *Air Power History,* Winter 2003, p. 21.
6. A common practice, exemplified by David Reynolds in *America, Empire of Liberty,* is to mention the first launch and immediately follow it with the second, in this case by saying: "The following month, the Soviet Union rubbed in its lead over the supposedly high-tech Americans by sending Sputnik II into orbit." The two, separate, events are then discussed for a further two and a half pages as one event.
7. Sourced from http://www.marshall.org/article.php?id=80526. Accessed October 12, 2012.

8. JS Lay, "National Security Council Report, Note by the Executive Secretary to the National Security Council on US Scientific Satellite Program," Washington, May 20, 1955, Special Assistant for National Security Affairs Records, DDE Library.
9. Press release: "Plans for Construction of Earth Satellite Vehicle Announced," July 29, 1955, Box 624, Official File, DDE Records as President, DDE Library.
10. Press Release: "Statement by White House Press Secretary, James C. Hagerty on Earth-Circling Satellites as Part of IGY program," July 29, 1955, Box 624, Official File, DDE Records as President, DDE Library.
11. Harford, *Korolev*, p. 123.
12. Information sourced from International Astronautical Federation, http://67-20-84-59.hostmonster.com/index.html?title=IAC1955. Accessed on 10 June 2014. The sixth International Astronautical Congress was held in Copenhagen and on August 2nd, news reached the participants of the US satellite announcement. Sedov responded the same day in a news conference held at the Soviet Embassy.
13. Harford, *Korolev*, p. 126.
14. AL Blakeslee, Associated Press science reporter, *The Courier News*, Blytheville, Arkansas, Jan 1, 1957, p. 5.
15. AL Blakeslee, "US Prepares to Probe Secrets of Outer Space," Associated Press, syndicated article, May 27, 1957.
16. VA Kotelnikov, "Soviet Satellites in the IGY," International Astronautical Federation Conference, Boulder, Colorado, August 27, 1957.
17. International News Service syndicated article, (unattributed) "Manned Satellites Hinted by Moscow," *Corsicana Daily Sun*, August 31, 1957, pp. 1 and 10.
18. Ellender travelled to Moscow three times from 1955–1957 and returned following an interview with Khrushchev saying: "The Russian people are better off today under communism than they have ever been. As much as I abhor communism, and as much as I dislike to admit that, it is the truth." TA Becnel, *Senator Allen Ellender of Louisiana*. (Baton Rouge, 1995), p. 203.
19. Unattributed, "Russ Satellites to Bare Space Satellite Plans," International News Service (INS) newswire, New York, September 30, 1957.
20. The INS "Russ Scientists . . ." piece was sourced from the *Pasadena Independent* newspaper, Pasadena, California, September 30, 1957, p. 4.
21. *Chicago Daily Tribune*, Chicago, October 6, 1957—Section 1 leader article, p. 2.
22. *New York Times*, October 6, 1957, p. 1.
23. *The Mansfield News-Journal*, Mansfield, Ohio, October 5, 1957, p 1.
24. Ibid.
25. *The Sun Telegram*, San Bernardino, California, October 6, 1957, pp. 1–2.
26. HM Schmeck Jr., "Moscow Predicts Satellite Will Stay Aloft for a Long Time," *New York Times*, October 10, 1957, p. 1.
27. *New York Times*, October 20, 1957, Section R, p. 22.
28. *New York Times*, November 1, 1957, Review Section, p. 2.
29. *Milwaukee Sentinel*, October 5, 1957, Section 1, p. 1.
30. "Army Tries Its Misguided Missile Man," *Life* magazine, July 16, 1957, p. 16. The court martial of Colonel John C. Nickerson, Jr. took place in June 1957. Nickerson pleaded guilty to 15 counts of breaching Army security regulations. The charges were based on his release of a document containing defense secrets, along with a personal memorandum highly critical of the Secretary of Defense's decision to limit the Army's missile and rocket R&D role to ranges of 200 miles or less. His legal team argued that his motive was to get permission for the Army to maintain operational control of the Jupiter missile system being developed by ABMA. On June 29, 1957, the General Court Martial board suspended Nickerson from his rank for one year, fined him $100 pay per month for 15 months, and issued an official reprimand. The personal memorandum, referred to in the Scripps-Howard news story was discussed in court in the presence of 71 reporters.
31. JG Lucas and D Preston, Scripps-Howard staff writers, Washington, DC syndicated feature sourced from *El Paso Herald Post*, October 9, 1957, p. 1.

32. In 1950, television penetration of US households was only 9.0 percent. By 1955, it had reached 64.5 percent. By 1965 it reached 92.6 percent. Sourced from US Television bureau of Advertising figures, 2010. In 1957, there were 1,755 daily newspapers in the United States, with a combined daily circulation of 57,805,000.

33. N Frazier, *William Randolph Hearst, Modern Media Tycoon* (Woodbridge, CT, 2001).

34. G Baldasty, *EW Scripps and the Business of Newspapers* (Urbana, IL, 1999).

35. *Athens Limestone Democrat,* Athens, Alabama, October 8, 1957, pp. 1 and 8.

36. *Pravda,* the Soviet daily newspaper, printed the TASS news announcement in full on page 1, but down-page, October 5, 1957.

37. *Chicago Daily Tribune,* October 6, 1957, Section 1, p.2.

38. Hall and Shayler, *Rocket Men,* p. 60.

39. The Soviets released a number of journal articles including Mikhailov's "On the Observation of the Artificial Satellite" published in the *USSR Astronomical Journal,* Vol 43, no. 3, p. 313—Moscow, September 1957. This was freely available to western scientists and, indeed, had been translated by US Intelligence.

40. S. Khrushchev, *Nikita Khrushchev and Creation of a Superpower* (University Park, PA, 2000) p.156.

41. On August 5, 1957, a U-2 mission was launched from Lahore, Pakistan, which captured the first pictures of the new Baikonur Cosmodrome near Tyuratam—see SD Schultz, "Why Gambit and Hexagon? U.S. National Security and the Geopolitical Setting, 1957–1960," in *National Reconnaissance,* issue 2012 U1, p. 3.

42. W Lippmann, *Public Opinion,* first published 1922 (Greenback Books edition, New York, 2010).

43. Lyndon Johnson's political use of the space issue is discussed in detail in chapter 4.

44. By 1957, Eisenhower's Defense budget cuts, aimed squarely at reducing conventional forces were beginning to bite. While Strategic Air Command was growing and defense contractors were beginning to see opportunities for more technologically-driven military solutions, the overall DoD budget had been cut by 20 percent from its 1952 high to $45.4 billion in FY1957. The defense lobby was the best organized lobby in Washington, working closely with members of both congressional chambers and the Pentagon, and becoming the first professionally organized lobby, paid for by the major defense contractors.

45. Both ICBM/IRBM development and continental defense systems had been prioritized as a result of the TCP recommendations in February 1955.

46. Magazines such as *Aviation Week* and its newspaper partner *Aviation Daily* acted as national mouthpieces for the aerospace industry since they were totally reliant on the industry for their survival and profit. Their "editorial" was largely created in the PR departments of the major corporations. While the corporate PR teams did not have quite the same relationship with more general national news magazines, they delivered a heavy advertising spend with full page advertising, often featuring "expert" significantly biased advertorial, was carried in the major news weeklies such as *Time* and *Newsweek* almost every week.

47. Daily News Conferences, October 5 and 7, 1957, Box 49, News Conferences, Hagerty Collection , DDE Library.

48. RH Ferrell, *Hagerty,* p. xlv.

49. Diary records October 4–8, 1957, Box 9, October 4 1957—January 14 1958, Diary Series, Ann Whitman File, DDE Files, DDE Library.

50. "Statement by the President Summarizing Facts in the Development of an Earth Satellite by the United States," October 9, 1957, document 211, The American Presidency Project, http://www.presidency.ucsb.edu/ws/index.php?pid=10925&st=&st1. Accessed September 9, 2014.

51. Killian, *Sputnik,* pp. 12–13.

52. D Bronk, "Science Advice in the White House," *Science,* 11 October 1974, no. 186, pp. 116–21.

53. W McDougall, *The Heavens and the Earth*, p. 186
54. "Memorandum of Conference with President Eisenhower on October 8 1957," A Goodpaster, 9 October 1957, item from DDE-EPRES Collection, DDE Papers as President of the United States 1953–1961, DDE Library.
55. Nathan Pusey was the president of Harvard and a member of the Science Advisory Panel.
56. *Chicago Daily Tribune*, October 6, 1957, p. 1.
57. Attributed to trade adviser Clarence Randall by RA Divine in *The Sputnik Challenge* (Oxford 1993), p. xiv.
58. Anon, UP syndicated story, Washington, DC, October 5, 1957. Sourced from *Miami Daily News-Record*, Miami, Oklahoma, October 6, 1957, pages one and three. Those aides showing concern are not named, although the article stated that "scientists," with no mention of who they were or their relationship to the president, said that the size of the satellite suggested Russia had not been "bluffing when it recently claimed development of an intercontinental ballistic missile." Meanwhile, equally unnamed sources at the State Department "feared that Russia may have scored its greatest propaganda victory." P. 1 for both quotes.
59. Anon, UP syndicated article, Washington, DC, 8 October 1957, published as "Russ Do Not Have Big Missile: Wilson" sourced from *Idaho State Journal*, Pocatello, Idaho, October 8, 1957, p. 1.
60. Diary records October 9, 1957, Box 9, October 4, 1957—January 14, 1958, Diary Series, Ann Whitman File, DDE Files, DDE Library.
61. J Hagerty, drafts of News Conference Statement on Sputnik, October 9, 1957, News conference file, James C Hagerty Papers, Organization and Early History of NASA 1957–1961, DDE Library.
62. "The President's News Conference," October 9, 1957, Document 210, The American Presidency Project, http://www.presidency.ucsb.edu/ws/?pid=10924. Accessed September 9, 2014.
63. Ibid.
64. Ibid.
65. Diary Records, October 9, 1957, Box 9, October 4, 1957—January 14, 1958, DDE Files, Diary Series, Ann Whitman File, DDE Library.
66. "Memorandum of Conference with President Eisenhower on October 9 1957," A Goodpaster, item from DDE-EPRES Collection, DDE Papers as President of the United States 1953–1961, DDE Library.
67. Unattributed, "See Satellite over the US," *Kansas City Star*, Kansas City, Missouri, October 10, 1957, p. 1.
68. *Kansas City Star*, Kansas City, Missouri, October 11, 1957.
69. Ibid., p. 1–4.
70. Interview sourced from C-Span Video Library: http://www.c-spanvideo.org/program/288845-1. Accessed on Jan 31, 2013.
71. General George Churchill Kenney, Official Air Force Biography, US Air Force, http://www.af.mil/AboutUs/Biographies/Display/tabid/225/Article/104770/general-george-churchill-kenney.aspx. Accessed on Jan 31, 2013.
72. The Nation, *Time Magazine*, Volume 70, no. 16, October 14, 1957, p. 22.
73. "Sputnik the Result of Soviet Science System," *Kansas City Star*, October 20, 1957, Section F, pp. 1–2.
74. E Lebreton, Associated Press, "Reds Exploit Sputnik," syndicated article, October 19, 1957, Washington, DC.
75. Presidential News Conferences, October 9 and 31, 1957, Box 63, Hagerty Collection, Presidential News Conferences, DDE Library.
76. Ibid.
77. Ibid.
78. Johnson's response to Sputnik is heavily quoted and forms a cornerstone to the traditional interpretation of the Soviet satellite launch. T Heppenheimer references it in *Countdown*, (Wiley, 1997) p.124, while Roger Launius quotes it (unreferenced) in RD

Launius, "Sputnik and the Origins of the Space Age" in NASA History Division's http://history.nasa.gov/sputnik/sputorig.html. Accessed June 13, 2014. It originally appeared in LB Johnson, *The Vantage Point*, (New York, 1971), p. 283.

79. Symington used the term when addressing the Hannibal Kiwanis Club on November 4, 1957. Sourced from *Schenectady Gazette*, November 5, 1957, p. 9.

80. "Journal's Opinion," *Salina Journal*, October 9, 1957, p. 2.

81. "Memorandum for Mr McElroy, Secretary of Defense, from OM Gale Special Assistant," April 14, 1958. Sourced from National Digital Security Archive, accessed December 6, 2012.

82. Ibid.

83. RD Launius, "Sputnik and the Origins of the Space Age," NASA History Office, (Washington, DC, 2005), http://history.nasa.gov/sputnik/sputorig.html. Accessed June 13, 2014.

84. Launius and McCurdy, *Myth of Presidential Power*, p. 27.

85. DDE, National networks TV address, November 7, 1957.

86. "A Time of Danger," *Time*, November 11, 1957, p. 23.

87. "Sputnik II," *New Republic*, November 11, 1957, p. 3. The *New Republic* was rather more conservative in its political leaning than *Time*.

88. K Gatland, "Russia's Second Satellite," published in *Spaceflight*, vol 1, no. 6, January 1958, pp. 204–5.

89. *Life* magazine, (New York), November 18, 1957, p.1, http://books.google.co.uk/books?id=vVYEAAAAMBAJ&printsec=frontcover&source=gbs_ge_summary_r&cad=0#v=onepage&q&f=true. Accessed May 13, 2013.

90. Ibid., p. 39.

91. "Our Future Security," speech delivered by President Eisenhower in Oklahoma City, November 13, 1957, http://www.eisenhower.archives.gov/research/online_documents/sputnik.html. Accessed September 14, 2014

92. Drafts of "Our Future Security," Boxes 1–6, series 1, Arthur Larson Papers, DDE Library.

93. Ibid.

94. Smith, *War and Peace*, p. 731.

95. *Chicago Daily Tribune*, Chicago, November 4, 1957, section 1, p. 1.

96. Ibid., p. 3

97. Ibid., p. 2

98. *Chicago Daily Tribune*, Chicago, November 5, 1957, section 1, p. 1.

99. Ibid., p. 1.

100. *Chicago Daily Tribune*, Chicago, November 7, 1957, section 1, pp. 1 and 3.

101. *Atlanta Constitution*, Atlanta, November 8, 1957, Section A, p. 18.

102. Sheehan, *A Fiery Peace*, p. 357.

103. The Atlas "A," produced by the Convair Division of General Dynamics, was the shortest (by 9 inches) and lightest of the Atlas missiles.

104. Schlesinger, *Imperial Presidency*, p. 159.

105. Richard G. Hewlett and Jack M. Holl, "Atoms for Peace and War: The Eisenhower Administration and the Atomic Energy Commission," *A History of the Atomic Energy Commission*, Volume III (Berkeley and Los Angeles, CA, 1989) p. 465.

106. R Sprague, Oral History interview, "War and Peace in the Nuclear Age," tapes c03005–c03007, 1986, WGBH Media Library, http://openvault.wgbh.org/catalog/wpna-dfe326-interview-with-robert-sprague-1986. Accessed September 14, 2014.

107. A Wohlstetter, *Protecting the US' Power to Strike Back in the 1950s and 1960s*. Sourced from albertwolstetter.com—http://www.albertwohlstetter.com/writings/19560901-AW-EtAl-R290.pdf. Wohlstetter is, of course, seen by many as the father of modern neo-conservatism. Accessed April 14, 2014.

108. Nitze had been at the heart of the lobby for increased defense spending under Truman. Taking over from Kennan, as Head of the State Department's Policy Planning staff, he had been largely responsible, along with Dean Acheson, for the national security review that led to NSC-68 and the tripling of US spending on conventional

and nuclear arms between its publication in 1950 and Eisenhower entering the White House in 1953 and had worked closely with the Committee for the Present Danger, founded to promote the swiftest implementation of NSC-68. The NSC-68 paper "United States Objectives and Programs for National Security," April 14, 1950, was published in Foreign Relations, 1950, volume I, pp. 234–92.

109. "Deterrence and Survival in the Nuclear Age," Security Resources Panel of the Science Advisory Committee, Washington, DC, November 7, 1957, http://www2.gwu.edu/~nsarchiv/NSAEBB/NSAEBB139/nitze02.pdf. Accessed September 19, 2014.

110. Ibid., pp. 4–5.

111. Ibid., pp. 11–14.

112. "Memorandum of Discussion at the 343d Meeting of the National Security Council, Washington, November 7, 1957," prepared by SE Gleason, November 8, 1957, NSC Records, Whitman File, DDE Library.

113. Ibid.

114. Ibid.

115. C Roberts, "Secret Report Sees US in Grave Peril" *Washington Post*, December 20, 1957, Section A1, p. 6.

116. Killian, *Sputnik*, p. 100.

117. "Memorandum of Discussion of the 350th meeting of the NSC, January 6, 1958." Drafted by S Gleason, Jan. 7, 1958, NSC Records, Whitman File, DDE Library.

118. Mieczkowski, *Sputnik Moment*, p. 292.

119. Ibid., p. 290.

120. DDE Presidential News Conference, October 9, 1957.

FOUR

From Confrontation to Legislation

Eisenhower's second term in the White House is too easily dismissed as a period when nothing happened. Ike was the "caretaker-president," allowing John Foster Dulles to shape and deliver his major policy relating to the Soviet Union.[1] While revisions to this caricature began as early as the 1980s, the long-outdated view still lingers. Therefore, it is important to take a look at the progress of the executive as Eisenhower and his advisers moved on from the first US ICBM tests and the ending of the IGY towards the implementation of the first US space policy. In doing so, what emerges is an effective case study demonstrating that far from being a caretaker, Eisenhower was still very much an engaged leader, developing and driving policy in support of his twin aims of enhanced national security and a balanced budget.

The development of space policy captures Eisenhower in microcosm: not micro-managing every operational decision, but working through expert "Helping Hands" to achieve his policy goals. The period covered here is from the end of the supposed furor of the Sputnik Fall, through to the enacting of legislation in July 1958. In reassessing the actions taken by key actors throughout the process, it focuses first on Lyndon Johnson, traditionally seen as the "founding father" of space legislation, before detailing the interaction specifically of Eisenhower and his PSAC advisers in moving from a military-led proposal for space development, to a civilianized solution that took the outcomes of both the IGY and ICBM/IRBM development programs, and shaped an effective way forward for the United States' action in space.

LBJ AND SPUTNIK

Following the launch of Sputnik 2, Senate majority leader, Lyndon Johnson, used his opening speech in the first session of the Senate Preparedness Subcommittee's post-Sputnik hearings to "ask the people in charge to tell us . . . how we can regain the leadership."[2] Over the next two months, Johnson attempted to expose the administration's perceived flaws in developing a credible missile and space program while he sought to present himself as the country's leading advocate for space exploration. Indeed, the interpretation of Johnson as a crucial player in the creation of NASA has been a staple of scholarly interpretation throughout the period of Eisenhower revisionism. In *Presidential Studies Quarterly* in 1994 for instance, Thomas Gaskin recounted Johnson's interaction with Eisenhower on foreign policy across Eisenhower's second presidential term. Gaskin delivered the standard interpretation for Johnson's influence writing:

> In the ensuing months [following his first round of Senate hearings, discussed below], Johnson got the Senate to create a Special Committee on Space and Aeronautics which he chaired. From this committee he helped shape the Eisenhower legislation that established the National Aeronautics and Space Administration (NASA), insured that it was under civilian control, and forced Eisenhower to accept legislation creating a Space Council which would set policy.[3]

Gaskin makes the common assumption that Johnson was at the center of the process to create NASA and a powerful mover in shaping the organization's purpose and governance. However, that assumption is questionable.

As owner of a number of TV and radio stations across Texas through Texas Broadcasting (later the LBJ Company) Johnson was certainly extremely media-aware. As such, he saw the issue of space policy as a means to gain prominence via the media. With the 1960 presidential race beginning to shape up, he also had an issue to use as a means to gain support for his candidacy for the Democratic presidential nomination. *Time* reported him addressing the 1958 Democratic Caucus saying:

> Our national potential exceeds our national performance. Our science and technology has been, for some time, capable of many of the achievements displayed thus far by Soviet science. That the Soviet achievements are tangible and visible, while ours are not, is a result of policy decisions made within the governments of the respective nations. The evaluation of the importance of the control of outer space made by us has not been based primarily on the judgment of men most qualified to make such an appraisal. Our decisions . . . have been made within the framework of the Government's annual budget. This control has, again and again, appeared and reappeared as the prime limitation upon our scientific advancement. . . . What should be our goal? If, out

> in space, there is the ultimate position—from which total control of the earth may be exercised—then our national goal and the goal of all free men must be to win and hold that position.[4]

Yet one must consider what he said in 1958. He was advocating increased spending on "tangible and visible" space achievements, and his objective was "to win and hold" outer space. Both points were anathema to Eisenhower who had no interest in a space race or any kind of Cold War proxy contest in space. Nor had he any intention of spending a dollar more than he had to in order to ensure the national security of the United States. Of course, space was also a "Trojan Horse" which provided Johnson with the means to mount a wider attack on the administration, undermining Eisenhower's budget-led approach and, implicitly, criticizing his fitness to govern.

Most Johnson scholarship focusing on the early years of US space effort stresses the significance of his contribution to getting the United States into the "space race" both through his subcommittee's pricking of the president into action, and through the masterly way he drove the 1958 National Aeronautics and Space Act through Congress. In his biography *Lyndon B. Johnson—Portrait of a President*, Robert Dallek perhaps best articulated the standard line on Johnson's congressional management of the space issue. He asserted that the first phase of the Preparedness Subcommittee hearings from November 1957–January 1958:

> allowed Lyndon to identify himself as the country's leading congressional advocate of a stepped-up effort in space. He dominated the hearings, introducing witnesses, leading cross examinations and making himself the principal spokesman to the press . . . Lyndon was the architect of the new organisation.[5]

Dallek's take on Johnson's senatorial command over his subcommittee's hearings deploys fairly traditional themes by taking much of its force from characterizing Eisenhower as a passive non-player, at odds with the American people and especially its politicians. But this viewpoint—and more so, Johnson's impact in championing the NASA legislation—has come under particular scrutiny recently, with Robert Caro, taking a rather different line on just how influential Johnson actually was. The orthodox line has Johnson playing a very significant role shaping the arms-length independent agency to manage the United States' civil space program, leading the process to move the plans for NASA from executive desire to congressionally-approved legislation. Pushing through the kind of legislation Eisenhower favored in a Democrat-led Congress was never going to be easy. Eisenhower required allies-of-circumstance: politicians he could work with even if their aims were somewhat different from his own. Johnson fitted that profile and, following their work together on the Civil Rights Act of 1957, Johnson's championing of the Space and Aeronautics Act of 1958 has generally been regarded as an extension of this relationship—another opportunity for the leader of the Senate to project his importance on the national stage and to an international audience.[6] This was all grist to his intended challenge for the Democratic presidential nomination in 1960. However, Caro takes a rather more nuanced view of LBJ's relationship with space legislation and thus with the president over it. Yes, Johnson used the aftermath of the first Sputnik launch for his own political gain, but when the time came to steer the legislation founding NASA and setting the first formal US space policy through the Senate, he was much more the passenger than the driver. Caro wrote that following the launch of Sputnik, Johnson was actually a little slow off the mark and not the first Democrat to react to the Soviet satellite. Indeed, Democrat politicians, whose first reaction was to assume that "the Administration had squandered America's lead in missilery and that the nation had been caught unprepared,"[7] turned first not to Johnson to challenge the Administration, but to Senator Richard Russell, chairman of the Senate Armed Services Committee (ASC). But Russell felt that LBJ was better suited to investigating just how "prepared" the United States was to meet this new perceived Soviet threat. Indeed, Johnson led the ASC's Preparedness Subcommittee.

Johnson assured the administration, Defense Secretary Neil McElroy and Senate Republicans that the investigation would be non-partisan. Yet his key intention was to demonstrate his ownership of the missile/rocket/satellite issue and to use it as a bolster to his impending entry into the race for the presidency. He stage-managed the hearings impeccably: creating a precise order intended to create the maximum impact on the administration, the media, and therefore the wider American public. First, he called those scientists deemed likely to embarrass the administration—Teller first, soon followed by the nation's most recognized space figure: Werner von Braun. Then, as Caro put it: "came the Generals (Le May, Hyman, Rickover and Gavin) . . . to paint a disturbing picture of how an overly economy-conscious Administration had allowed its emphasis on a balanced budget to interfere with the nation's security."[8]

The Senate Preparedness Subcommittee met from November 25, 1957, to January 23, 1958, "on matters pertaining to outer space" and a summary of selected testimonies was used by Johnson's Space and Astronautics subcommittee as it began to debate the administration's proposed legislation in April.[9] In total, seventy-three witnesses appeared at the hearings, and their testimony totaled 2,376 pages.[10]

SENATE HEARINGS

On November 25, 1957, Teller was the first significant scientist to face the Preparedness Subcommittee. It was an ominous day since it was also the day when the president suffered his stroke. While he was active in the

White House within a week, the illness effectively took Eisenhower out of the public arena until the following January. Thus, as Eisenhower succumbed to the onset of old age (at least, according to traditionalist accounts) coupled with the dangers of a 1950s lifestyle (though an ex-smoker, he consumed more alcohol than one would consider wise for the leader of the free world today), at the other end of Pennsylvania Avenue, Johnson attempted to twist the political knife. Teller had been active in criticizing the executive in the wake of the Sputnik launches. Evan Thomas described Teller as "a reliable purveyor of gloom and doom"[11] and the self-styled "father of the H-bomb" did not disappoint, although one has to question whether he was the right person to be addressing the committee at this point. Teller was an expert on nuclear physics, not on satellites. Yet, in the more deferential 1950s, his views were sought and shared on all aspects of science.

> The Russians are definitely ahead in the ballistic missile and satellite fields. . . . Unless we get an engine with a large thrust, we will be behind in the general field of the control of outer space. . . . Control of outer space is as important, if not more important than the ballistic missile. If the Soviet Union should control over outer space with satellites before we do, this country will be in mortal danger . . . if the Russians chose, and they had the hydrogen warhead, they could put a hydrogen bomb on top of the Capitol.[12]

Teller's polemicizing was entirely without any basis in fact. He vastly overestimated the power of the Soviets and also their intent, while being entirely outside Eisenhower's circle of command (and therefore information) when it came to developing space and missile policy. Teller was an arch conservative, intensely hawkish and at odds with Eisenhower's moderate Republicanism. His testimony, which unsurprisingly gained significant coverage in the media, was focused on building bigger, more powerful missiles—something Eisenhower knew was not necessary. While an eminent physicist, he was neither an intelligence nor missile expert. Thus, his testimony was accorded rather more value than it was actually worth when it appeared in the newspapers the following day. The widely syndicated Roscoe Drummond column, which appeared in newspapers across the Midwest commented:

> In vital aspects of science, Dr. Teller—who has the credentials to speak and be heard—says we are about ten years behind. In ballistic missiles, the lag is about eighteen months.[13]

While Teller's remarks gained prominence, there were slightly different viewpoints expressed by witnesses in the first week of hearings. None, though, particularly redressed the balance in favor of the president.

Also on that first day of testimony, Dr. Vannevar Bush, who headed the Office of Scientific Research and Development and thus had been responsible for mobilizing the nation's scientists during World War II (including Jim Killian) wanted another wave of mobilization, this time to:

> make sure that from here on in, our missile and satellite development programs move on smoothly and effectively at maximum speed. . . . We should provide for unified military planning in everything we do so that future developments will be in order from the outset. I would not recommend at this time a civilian organization outside the Department of Defense to carry on military development. Co-ordination can be done better within the Department of Defense.[14]

Bush was concerned at the internecine rivalry that still stalked the Pentagon and each service's development programs. His sentiment made sense, but no doubt appeared another criticism of the executive to the watching media, their viewers, listeners, and readers.

General James Doolittle testified the following day. As a war hero best remembered for leading the B25 bombing raid on Japan in 1942 as well as his key role in the bombing offensive over Germany,[15] still a towering figure in the American imagination, his words carried weight—more so as he had been invited to testify as chairman of the National Advisory Committee for Aeronautics (the NACA).[16] His essential line was simple: "We need more basic research."[17] Taken with the headline comments of the previous day, his testimony began to paint a picture of a president asleep at the wheel; who had allowed the United States to fall behind the Soviets and thus put the nation at risk. While Eisenhower was far from asleep, he was most certainly incapacitated due to his stroke as Johnson invited witnesses to testify before the committee. Therefore, the president's early defense fell to his appointees in the Department of Defense, his newly-appointed secretary, the businessman-turned-politician, Neil McElroy[18] and then Assistant Secretary Don Quarles, who was rather more engaged on the inside story of the development of the United States' missile, and therefore rocket capability.

McElroy's testimony of November 27 was short and to the point. But under questioning, he failed, once again, to uphold the Administration's line. First, supporting the president, McElroy stated: "Dr. Killian can make a real contribution if he improves coordination in various research activities in government, such as National Science Foundation, NACA and Atomic Energy Commission." This was helpful in alerting the US public that Killian was much more than a "missile czar." But if Eisenhower hoped that McElroy could keep to the line first played out from the White House via John Foster Dulles and Jim Hagerty on October 5 that Sputnik had no impact on US actions either in missiles or satellites, then that hope was rapidly dashed under cross-examination. "Had we known the Sputniks were to be launched we would have done things differently. Sputnik surprised me," he said.[19]

He also said that if he had been privy to the intelligence knowledge, he would not have been so surprised. It was a brief flash of a chink in the armor, which, surprisingly was not picked up either by the committee or the media. McElroy seemed to suggest that the president had access to intelligence information regarding the launch of Soviet satellites. Yet the president or his spokesman had given no indication of that in any public comment. However, McElroy neglected to mention that boasts of an imminent Soviet satellite launch were being made quite publicly by Sedov at the end of September.

Don Quarles, speaking on the same day, was much more rigorous in his justification for the pace and priority of American action, and attempted to defuse some of the more polemic interpretation of the Sputnik satellites:

> The satellites do not prove the Russians are more advanced in rocketry than we are. Even if they did it would be a minor rather than a major factor in the near term balance of military power. . . . The military significance of launching a satellite can be exaggerated by failing to take account of all the other factors that do tend to make such a test of this time relatively less important than it was represented to be. [Other factors such as bomber retaliatory power] . . . does not entitle one to believe that they have, either have had at that time or will have in the near future, a striking power based on this kind of development.[20]

Quarles questioned the assumption, prompted by Johnson, that a satellite equated directly with the ability to launch a nuclear missile strike. To that extent, his statement was beneficial to Eisenhower. But prestige mattered as much as the reality of national security, and for the US public, the questioning of Quarles gave further credence to the belief that America could have launched the first satellite, and that the blame for not doing so lay at the president's door. Quarles noted that the United States could have launched its satellite ahead of the Soviets. "Had we started early enough, I think there is no doubt, we could have launched our satellite ahead of theirs."[21] He explained the subsequent US course of action by saying that in 1955, when the IGY was being prepared, the "best judgement" was to separate the civilian satellite from the military missile program and that the Navy was best placed to handle the job through Vanguard. This would ensure that IGY work would not interfere with the top priority of the ballistic missile program. However, hindsight forced him to agree that the "satellite lag" was in part caused by the lack of a unified program—a criticism of DoD and therefore by implication, the president. However, he noted that the civilian scientists were also partly to blame through their desire to get more and more instrumentation into the satellites. He acknowledged that the Army could have orbited a satellite using the Jupiter booster, but that this would have required additional investment and that was not favored in 1955/1956 by the public. Quarles had spread the blame for "failure" far and wide—with the president far from blameless.

Lieutenant General James M Gavin, who had come to prominence as the deputy division commander of the 82nd Airborne who had parachuted into France as the advance force in the very early hours of D-Day and had become a national hero, led the military fusillade describing Sputnik as: "Perhaps the most significant thing of our times."[22] While Gavin saw sense in creating a civilian organization to manage space exploration, he was in the military minority. His colleague from the Air Force, Clarence Irvine, deputy chief of staff for materiel, was adamant that further funding for research, development, and deployment should reside with the US Air Force. His most memorable response to the questioning was to say: "We do not need any more commissions, czars or organizations."[23] In between Gavin and Irvine came the army artillery general, John Medaris, and Wernher Von Braun. Medaris, reciting the chronology of the Jupiter-C development, was equally transparent in his case-setting for winning the development of satellites for the Army. His view was that all missile development should be the responsibility of the Army—since a missile was merely an extension of an artillery piece. He pointed out that the Army had long had the capability to launch a satellite, having successfully tested the satellite configuration in September 1956. Preempting Von Braun's testimony which was to follow, he noted that the missing element in missile (and therefore rocket) development was the "big thrust engine." He did not see satellites and missiles being an either/or decision, and nor did he see the need to prioritize one over the other. "The priority [for the United States] should be on the attainment of a space capability at the earliest possible date. Satellites and ballistic missiles have many basic techniques that cannot be separated. Divorcement [sic] of the two impedes both."[24]

Von Braun, the German-born engineer who had become the public face of US rocketry through his TV exposure and numerous articles in *Life, Colliers,* and other widely-read publications, followed his military manager. His testimony was timely in so far as it came eight days after the failure of the Vanguard launch, when the rocket blew up just feet above the Cape Canaveral launch pad, and ten days after his *Mars* space show had been aired on the ABC network.[25] His line was trenchant, predictable, and self-serving: "The Russians are definitely ahead in the ballistic missile and satellite fields," he said.[26] "Unless we get an engine with a large thrust, we will be behind in the general field of the control of outer space. . . . If we want to establish control of outer space by manned vehicles we will need large engines. I believe the control of outer space is more important than the ballistic missile." This was very much against Eisenhower sentiment, since "control" of space implied some element of battle, and the one thing that the president was absolutely adamant on was the peaceful exploration of space.

Eisenhower was a long-term, strategic actor. His philosophy was to wage peace, and thus only fight battles as a last resort, and only when he knew he had the capability to win. Outer space was a potential battlefield that Eisenhower's forces were not going to be in a position to control. As early as 1947, in a West Point commencement speech, Eisenhower had said: "War is mankind's most tragic and stupid folly; to seek or advise its deliberate provocation is a black crime against all men. . . . For Americans, only a threat to our way of life justifies resort to conflict"[27] Therefore, in his way of thinking, space, where there was no actual threat to the way of life of US citizens was never going to be a battlefield. That made Von Braun's later statements on the capabilities of artificial satellites largely redundant, although he was prescient (if out of the information loop) in his understanding of the potential for reconnaissance satellites. Portentously, Von Braun agreed that Sputnik proved that the USSR had the means of sending a hydrogen warhead anywhere in the world and that a Sputnik-type vehicle gave them orbiting bombing capability. Within months, Ed Purcell's paper was to prove that notion to be far-fetched. However, Von Braun was right in saying that a satellite the size of Sputnik 2 "would be entirely capable of carrying a combination of optical and television equipment to use as a powerful reconnaissance instrument."[28]

By 1960, Eisenhower had deployed such satellites. Von Braun, of course, was using the Capitol stage as a platform for advancing his own claims for a central role in the US space program. He reiterated that his team was ready to: "get a man into space in five years and build a space station in 10." He asked for an annual budget of $1.5 billion to make this happen, and was open to control of such a program being handled either by the Department of Defense or a new agency. Notably, Von Braun stated that the ideal setup for the United States in space would be for the consolidation of space and military effort under one man. He was, it seems, too modest to suggest a name.

Von Braun was highly skilled in playing to the crowd—be they the assembled senators or the wider public reached through the media. Certainly the *Washington Post* regarded his testimony as important and appeared swayed by the force of his argument. The following day, it led with the headline: "All out push to dominate space urged."[29] Overall, Von Braun's skillful answering of Johnson's committeemen's questions earned positive reviews. The *New York Times* described his testimony a "hit."[30] It seems fair to say that his intervention was the apogee of the Johnson hearings. After Von Braun's bravura performance, the hearings became rather repetitive, with the repeated request for more money to be pumped into the DoD varied only in tone and dependent on which service the speaker was representing. Lieutenant General Donald Putt, the Air Force's military director of its Science Advisory Committee and deputy chief of staff for development, summed up the military mood saying:

"We need funds for research and development on future bombers, nuclear propulsion for aircraft, electronics, missiles, satellites, chemical propulsion and new fuels."[31] There were not too many areas of potential spending that he chose to leave out.

Johnson's committee broke for Christmas on December 17, and never really regained its energy in the New Year. That said, Johnson was still keen to emphasize the distinct gap between his thinking and that of the president, stepping out of the supposed bi-partisan environment of the committee hearings to do so. Just three weeks before the United States launched its own satellite, Johnson reflected the critical thinking around the fear of Soviet domination in space. "There is something more important than the ultimate weapon," Johnson declared when he addressed the Democratic Caucus. "That is the ultimate position—the position of total control over Earth that lies in outer space."[32] Johnson's sentiment exposed a radical difference in his thinking on space from the president's. Eisenhower saw good reason to exploit space militarily for communication, meteorology, and reconnaissance purposes, but did not rate space as likely to be highly important in future wars. Johnson, on the other hand, was an early advocate for the weaponization of space and certainly saw the military having command of all US space efforts.

The one testimony of note from the January committee hearings came from Nelson Rockefeller. He spoke on January 10, 1958, in his position as chairman of the President's Advisory Committee on Government Organization, a role he had held since 1953. Although a moderate Republican, Rockefeller, grandson of the founder of Standard Oil, was both independently extremely wealthy, and set to run for political office (he was elected governor of New York later in the year, and was a rival to Nixon's presidential run in 1960), and thus the stage offered by Johnson's committee gave him a national platform to reinforce his credentials and views with the American public. His answers to the committee's questions highlighted a difference in thinking from the president. "The Russians have given space a higher priority of effort. Our scientific effort should be intensified," he commented.[33] And although, even at this early stage in the process, he was well aware of Killian's PSAC drive to secure the civilian space agency through a reorganized NACA, he said: "Problems of where research in space should be done should be decided by the Secretary of the Department of Defense." He also stated that the decision on whether the new space agency should reside within the Pentagon or not should be decided not by the president, but by the secretary for defense. Such sentiment hardly helped the administration's case—but also rather pointedly showed that Rockefeller was not an Eisenhower insider in the evolution of the administration's first space policy.

WRITTEN EVIDENCE

In addition to the in-person hearings, the committee also reviewed written evidence, with weight being given to the American Rocket Society which had first written to Eisenhower on October 14 in the wake of Sputnik 1's launch and also to the Rocket and Satellite Research Panel, which had written directly to the committee. Their papers were combined as a joint proposal for a national program for space flight. The Rocket and Satellite Research Panel's signatories spanned the upper echelons of military research, supported by their key suppliers (Convair and General Electric), and included Von Braun from the Army Ballistic Missile Agency and Pickering and Stewart from the Jet Propulsion Laboratory. The American Rocket Society's report, signed by its president, George Sutton and James Van Allen whose Explorer satellite was soon to be America's first artificial satellite, focused a little less on money (the Rocket and Satellite Research Panel was looking for additions to the defense budget of no less than $10 billion), and was prescient in the program it presented. This focused initially on a series of unmanned scientific earth orbits, and both moon and Venus circumnavigation flights, followed by "Returnable, manned satellite in flight around the earth by 1961 or 62," and "Small inhabitable permanent satellites by 1963 [space stations]." Suggesting how exploration might develop later in the decade, the Society called for a "Manned expedition to the moon by 1 or 2 men by 1968; establishment of a permanent human [moon] base, if desired, by 1970" and "Fast manned reconnaissance flight to Mars and Venus, without landing, beginning 1972."[34]

While Johnson generated a wealth of news coverage up to and through the hearings, they dragged on too long and their findings were undercut by the successful orbiting of the first US satellite, Explorer 1. So where did that leave Johnson? It would seem that he was left with a short-term hot issue that did much to create the concept of the "missile gap"—but an issue that by February was losing heat to be replaced by the domestic worries of the onset of recession. Indeed, the launch of the Explorer satellite on January 31, 1958, did much to bring public opinion in the United States back in synch with Eisenhower's unwavering line that America had nothing to fear from the launch of Soviet satellites. Johnson's subcommittee report had seventeen rather general recommendations for strengthening military security which included advocating a national program for space exploration. However, there was very little by way of specifics, and it failed to include plans for a space agency, either military or civilian.

As public opinion shifted towards the president on space and missile issues, the media's appetite for Johnson's crisis-talk faded. As soon as that happened Johnson's interest switched to what really mattered to him: poverty, education, and economic opportunity. Caro has shown that

that although the bills to create a new Senate special committee on Space and Astronautics were introduced to the Senate by Johnson in February, they held no interest for him. More importantly, we will see next that the core legislation to create NASA was drafted not by the committee but by the Eisenhower administration, principally by Killian's PSAC team.

Johnson's Preparedness Investigation had generated a lot of heat but exponentially less light, providing mainly a platform for disgruntled scientists and budget-hungry military figures to take the opportunity to indulge in what Divine describes as "military sour grapes."[35] However, contrary to Caro's outline of his involvement, LBJ's moment in the sun was not yet over. He was able to keep at the forefront of the outer space debate as both the House and Senate debated proposed legislation covering both the United States' first formal civilian space program, and how that program would be administered. But, his role changed significantly in 1958, as the executive wrested control of the space issue back into the hands of the administration.

The testimony delivered at the hearings was important in so far as each piece of evidence was mulled over in the media and those portions that attacked the Eisenhower administration's policies have later been picked up and sustained by orthodox interpreters of the Sputnik "crisis." Certainly, Johnson's statements and the testimonies of the likes of Gavin and Von Braun, form cornerstones to the space race chronology. Yet they had no actual bearing on Eisenhower's parallel track. Killian was already at work assembling the team that would define and shape NASA and its role within the administration's space policy, while the Explorer satellite's launch, enabled by Von Braun's ABMA Jupiter rocket team, deflated his own polemic testimony from the previous November.

DAMAGE TO PRESIDENTIAL APPROVAL

Towards the end of 1957, Eisenhower's approval ratings had fallen from 79 to 57 percent—by far the lowest rating of his presidency.[36] But it was really only in the very early days in October that he had appeared out of step with the American public. Their view, short-term as it was, no doubt was heavily influenced by the agenda-setting media titles with a national reach such as the *New York Times, Washington Post,* and the news weeklies, which, albeit briefly, lined up against a perceived slow-moving presidency. Indeed, *Time* crowned Khrushchev its man of the year for his Sputnik successes.[37]

Eisenhower's stroke in late November hindered his ability to put the Sputnik satellites in true context. The messages he had begun to spread through his reassurance talks, first the national TV address on November 7, and second, the Oklahoma speech on November 13. Both had been well received. But his stroke meant he was unable to deliver the third planned talk. So, as Killian was making plans, out of sight of the media gaze, to draw together the threads of civilian and military rocket activity, Ike was rationing his appearances and involvement in day to day White House affairs and thus unable to contribute fully effectively to the shaping of a post-IGY military and civilian environment. The only publicity surrounding US space involvement was the daily reporting of Johnson's Senate hearings.

American prestige dropped further early in December with the first test of America's Vanguard rocket. This was to be launched from Cape Canaveral on December 6, 1957, in the full gaze of the media with an onlooking public audience of thousands and a TV audience of millions. For the scientists advising the president in Washington and for the engineers directly involved in the program, it was merely a staging point on the way to perfecting the Vanguard rocket. But particularly to a media cohort stung by Sputnik, this "launch" was all about catching up the Soviet space lead. However, "Test Vehicle 3" exploded four feet off the pad.[38] Thanks to live TV coverage, necessitated by Eisenhower's requirement that IGY activities be entirely open to public scrutiny, millions watched the unfolding debacle in amazement. The humiliation by the media was more international than domestic. In London, the *Daily Express* led next day with the headline: "US Calls It Kaputnik."[39] The *Daily Herald* was no less scathing with: "Oh What a Flopnik!"[40] The French were scornful, if opaque, with *Paris-Journal*'s "It seems there is a worm in the grapefruit."[41] Of course, the US regional dailies were not slow to report on the test failure. Their response was typified by the *Louisville Courier-Journal* which reported: "A shot may be heard around the world, but there are times when a dud is even louder."[42] At the United Nations, the tongue-in-cheek Soviets asked their American counterparts if the United States might wish to receive assistance under the Soviet program of foreign aid for technical assistance to backward nations.[43]

CREATING A DISTINCTION BETWEEN MILITARIZED AND CIVILIAN SPACE

While Eisenhower had been willing to let his three armed services compete to develop both satellites and missiles in the hope that competition would bring forth rapid technological advance, the effect had not been what he hoped for. However, much had been going on behind the scenes, not catalyzed by Sputnik, but more as the next aspect of a continuum that had begun at Bikini Atoll and was entering a natural next phase with the final months of the International Geophysical Year.

The most important aspect of Eisenhower's civilian space policy, developing NASA from a wealth of possible approaches to an entity bounded by legislation, occurred between the start of February 1958 and the end of March when Eisenhower instructed Killian to develop both objectives for the United States in space and the organization necessary to achieve them. Over this short period, a PSAC subcommittee, reporting to Killian but ably led by Nobel laureate, Ed Purcell, combined the task of setting the outline for the United States' first formal civilian outer space policy *and,* in parallel, defining a structure for bringing that policy to life. This combination proved a masterstroke. In so doing, it provided an insight into why Killian's relationship with Eisenhower was so effective.

The launch of the Explorer satellite had received what Whitman described as a "muted" reaction within the White House.[44] Yet this was not the whole truth. Despite being in his golfing cottage on the Augusta National Course, Eisenhower *had* taken an interest in the launch, and General Andy Goodpaster kept a record of a three-way telephone call between himself, the president, and Jim Hagerty over the course of the launch until a signal was picked up from the satellite denoting its safe orbit.[45] In the eyes of the mainstream American public, the launch of the Explorer satellite negated the Soviet lead in space. For the Army Ballistic Missile Agency headed by General Medaris the successful mission was a lever to be used to wrestle control of the forthcoming space program—while it also served to double the intensity of the Air Force's rivalry with the Army for the control of the space mandate.[46]

On February 4, PSAC instituted the Purcell Panel.[47] On February 21, a key participant in the panel, S. Paul Johnston, director of the Institute for Aeronautical Sciences expressed four organizational alternatives for a way forward. The four options, were: first, establish a new government agency. This would be costly and at odds with Eisenhower's strategy of limiting government spending. Second, assign the space program to the Atomic Energy Commission. This would be a popular decision with Congress, but the Panel noted, via Johnston, that the AEC had no experience in the space field, and space would be an unwanted distraction from its vital atomic energy roles. The third option was to establish the NACA as the controlling agency. Johnston stated: "[e]xtending [the NACA's] interests into space technology would seem to be a logical evolutionary step from its research activities of the past 40-odd years." The final option discussed was to assign space to the Advanced Research Projects Agency (ARPA) of the Defense Department. ARPA was created on February 7, 1958. "ARPA could take on the job with a minimum of additional legislation," wrote Johnston, "but military interests might outweigh the purely scientific and civil aspects."[48]

Initially Eisenhower's preference was for the military to maintain control of all rocketry, but this was primarily for budgetary reasons—he saw no reason to divert money for missile research into civilian space exploration.[49] But he had also maintained from the outset of the IGY that uses of space beyond military needs should not rest in the hands of the DoD and LA Minnich, assistant staff secretary, noted in his record of the meeting:

> The President's feeling was essentially a desire to avoid duplication, and priority for the moment would seem to rest with Defense because of the paramountcy of Defense aspects. However, the President thought that in regard to non-military aspects, Defense could be the operational agent, taking orders from some non-military scientific group.[50]

Killian, Herb York, and Kistiakowsky took part in this meeting at the White House on February 4 and Eisenhower met them again three days later. Across the meetings he reiterated to them that military space objectives had to have the highest priority, but in setting general capabilities for civilian space exploration, they had to resist competing pressures. What was left unsaid was where these pressures would come from though it can be assumed Eisenhower meant the branches of the military, the aerospace industry, and Congress. While Eisenhower had no interest in competing in any kind of a space or missile race with the Soviets, he was increasingly aware of the public impact of perceived Soviet successes. In his "Memorandum of Conference with the President" covering the February 4 meeting, Goodpaster wrote:

> The President stressed the importance of picking out the phases of activity in which we should undertake to compete with the Soviets, and to beat them. We should not try to excel in everything. He added that psychological as well as technical considerations are important—at times appearances are as significant as the reality, if not more so.[51]

Essentially, he was not opposed to "races" per se, but only wanted to compete in those of his choosing—again, reverting to his strategic intent of never fighting a battle unless he was sure he could win.

As Goodpaster noted, Killian took the president, Rockefeller, and Percival Brundage, director of the Bureau of the Budget, through the options for the new organization in a meeting on March 5.[52] Killian followed Eisenhower's wish to put military programs first—highlighting the role of the ARPA which had been established under DoD Directive 5105.15 on February 7, 1958.[53] Goodpaster recorded that Killian: "stressed the need to make use of existing facilities and competence, the limited scope of military space activity as presently foreseen, the need for a civilian agency to handle the civil aspects, with Department of Defense (ARPA) handling the defense aspects. He indicated the recommendation to use the National Advisory Committee for Aeronautics [NACA], substantially reconstituted and made responsible to Presidential direction."[54] In broadly accepting Killian's recommendation, Eisenhower's response considering the new civilian agency (referred to as NASA in the memorandum) was something of a feat of mental gymnastics. Goodpaster recorded that:

> It seems to him that military activity on space projects is acceptable in the area of application of knowledge. He feels certain, however, that discovery and research should be scientific, not military. He felt there is

> no problem of space activity (except ballistic weapons) that is not basically civilian, recognizing that application of findings may be made to serve military purposes.[55]

As Sean Kalic has recognized, Eisenhower was certainly prepared to *militarize* the use of space, but not to *weaponize* it, and drew a distinction between scientific exploration and discovery and more overtly military aspects that were rather less apparent to succeeding presidents.[56] However, it was the necessary distinction to bring NASA into being. It was not a distinction either the scientists or most certainly the military minds had drawn in Johnson's Preparedness Subcommittee hearings. It was, however, the distinction that Eisenhower had drawn in 1955 when prioritizing spending on military missiles, while creating a separate non-military budget in support of the IGY-related development of the Vanguard rocket and satellite by the Naval Research Laboratory.[57]

While even at this stage the president and Killian were in agreement on the organization of the new agency, the Atomic Energy Commission, DoD, and even both the Army and Air Force were still lobbying hard to be the prime player in the solution the Killian/Purcell team was planning. Whatever the solution was going to be, it would be a distinctly difficult balancing act. But Purcell managed this with panache by partnering the discussion on the governmental organization of civil space programs with the very elegant "Introduction to Outer Space" paper issued by the White House on March 26, 1958.[58] Killian and Purcell had first introduced the concepts of the paper in a briefing to the cabinet on March 14. The elegance of the paper is still apparent in a number of ways. First, though Purcell was a leading Harvard physicist and Nobel laureate, his paper is easily accessible to a lay audience. At the outset, it offered four reasons for the United States to explore outer space:

- The compelling urge for man to explore and to discover
- To use space to defend ourselves
- National prestige
- Scientific observation and experiment

Purcell's paper presented a pragmatic approach to space exploration suffused with both a latent excitement and scientific curiosity. Describing first the principles of orbital flight and rocket thrust, the paper offered an insight into sending unmanned craft to the moon, then to Mars and to Venus. It looked at the costs of scientific exploration and stated the benefit of scientific satellites for meteorology and communication. While focused on the civilian aspects of any future program, it did not shirk "The Military Applications of Space Technology." Here, the paper was at its most powerful, laying out the military program for satellite reconnaissance, communication, and meteorology. The text highlighted in particular the benefits of satellite reconnaissance.[59] This was a masterstroke of "hiding in plain sight." The Discoverer/CORONA program was well into development and the paper effectively summarized what this "black" satellite program would do. The brief but persuasive summary no doubt owed much to the presence on the committee of Polaroid's Din Land—the chief scientific innovator behind CORONA's powerful cameras. Allied to promoting the benefits of space reconnaissance, the paper was scathing in its dismissal of space as a "future theatre of war."[60] Purcell's text noted that there have been such suggestions as "satellite bombers and military bases on the moon." However, he noted: "even the most sober proposals do not hold up well . . . most of these schemes . . . appear to be clumsy and ineffective ways of doing a job."[61] In just a few sentences, the PSAC panel entirely undermined much of the reasoning for the armed forces/DoD taking control of the proposed new organization for the management of the civilian outer space program.

Killian, Purcell, and Herb York, the Berkeley Physicist who became ARPA's first chief scientist, took the cabinet through their "Introduction to Outer Space" paper on March 14, 1958.[62] In his memoir, Killian noted that "Eisenhower read it with enthusiasm and decided to use it to kick off a press conference."[63] At the press conference, the president asked the media to give the report the widest possible distribution. It gained wide and positive global coverage because, as Killian put it, "It was both a policy statement for the US government and an absorbing essay to read."[64] The latter point is down to both good planning and good crafting. Purcell and Land combined to draft the paper. Francis Bello, a journalist who had worked on *Fortune* magazine gave it journalistic flow and impact. Killian added authority with the introduction and conclusion. But the real clincher was an added statement from Eisenhower which prefaced the printed edition. He wrote: "This is not science fiction. This is a sober, realistic presentation prepared by leading scientists."[65] It is easy to detect the implied criticism of Wernher Von Braun, the Army Ballistic Missile Agency's chief engineer who had been lobbying hard, not least through a series of TV specials produced by Walt Disney, to win the new space agency for the Army.[66]

The solution proposed was the very opposite of Irving Janis's concept of groupthink, Janis's idea of a desire for consensus among groups working on a project resulting in the delivery of an irrational or dysfunctional—definitely suboptimal—solution.[67] Yet here was the opposite. Eisenhower initially favored a military solution; his expert scientific advisers came up with a proposition that maintained the high priority for military programs, yet placed the scientific exploration elements of the US space program in civilian hands. Eisenhower was convinced of the validity and moved his position accordingly while actually gaining greater control of the non-military program as an arms-length executive agency, and continuing to manage the missile and reconnaissance satellite programs within the set boundaries of his New Look policy.

FROM THE NACA TO NASA—DRAFTING THE LEGISLATION

Once the Killian/Purcell "Introduction to Outer Space" paper had been discussed in the NSC, and the PSAC team had presented their recommendation on the best way to organize such a program, the NACA emerged as the clear favorite to take on the management of the new civilian space program. This would require budgetary support; the agreement of the NACA's management; approval from the competing factions of the DoD and finally, approval by Eisenhower. This was an area where Killian thrived. He was not a scientist. He worked with scientists in both his academic role and throughout his succession of government appointments. But his true strength was as a shrewd administrator and master logistician. His skill lay in knowing who to win over for any decision, how to get to them and how to win them over. The trail of his meetings in February and early March show him mimicking a prime Eisenhower tactic: getting to the key players in any decision-making process first on a one-to-one basis before any major decision-making meetings took place. Killian met with both McElroy and Quarles in DoD, bringing them on board (though his memoir notes that they did not seem to fully realize how much operational responsibility would be devolved to NASA).[68] He also stayed close to the Bureau of the Budget (BoB) which would be responsible for the formal proposal reconstituting the NACA.

Paul Johnston's proposal to widen the NACA's remit appeared to chime particularly well with the BoB since its remit was to restrict the growing complexity of government. Prior to PSAC working with BoB to draft a memorandum for the president's approval, Killian also met the prime movers in the committee on government organization. This could, potentially, have been a tricky meeting since Rockefeller had previously spoken publically, not least in Johnson's hearings, of his wish for the entire space program to come under the auspices of the military, a view he had held since the initial discussions on the IGY back in 1955.[69] However, faced with the wealth of evidence provided by Johnston, Purcell, and the rest of Killian's PSAC team, Rockefeller was won over to the NACA proposal and jointly signed the memorandum to the president with the BoB's head, Percival Brundage. Still though, Killian was keen to ensure no surprises. Before formally presenting the memorandum for approval, he met Eisenhower privately one-to-one to take him through the detail of the key recommendations. Then, on March 5, at the request of Rockefeller and Brundage, Killian formally presented the memorandum to the president. The "no surprises" strategy paid off, and Eisenhower requested Brundage, with Killian's assistance, to proceed formally in drafting a bill for Congress.[70]

RACING CONGRESS

Killian recollected that Eisenhower wanted to get the space bill put before Congress before the Easter recess.[71] That required a draft to be ready for circulation by March 27 (Eisenhower had approved the drafting of a bill only on March 5), and for departments to turn round their comments by the following Monday. For a bill with significant potential impact on the DoD, this gave very little time for comment or room to maneuver. Indeed, during the Senate hearings on the bill, Lyndon Johnson somewhat acidly stated that the draft had been "whizzed through the Pentagon on a motorcycle."[72] This comment revealed Johnson's ambiguous status. Far from being the driver of the legislation as claimed by scholars such as Dallek, his function was rather closer to being a messenger. Certainly, this was a bill unlike many others. It was born not in Congress, but in the administration. And, as a significant rarity, it had initially been crafted by scientists. It is apparent that the process of drafting the bill had been underway for some time, based on the assumption that the proposal to expand the role of the NACA would be accepted. In fact, just how the idea behind the bill became the instrument placed before Congress provides a significant insight into the workings of Washington lawmakers at the time.

Much of that insight comes from Paul Dembling, general counsel for the NACA in the post-Sputnik period. He recalled the competition between the Army, Air Force, Atomic Energy Commission, and the NACA as to which organization would be designated to lead the US efforts in space.

By the start of 1958, the NACA was lobbying hard and one of its tools was, noted Dembling, to "draft a piece of legislation that we would be satisfied with if we got the choice of being the agency to handle the space program."[73] Hugh Dryden, director of the NACA gave Dembling the space to work on the legislation, while Dryden's number two, John Victory gave him the insight that: "Washington operates on the first draft it gets . . . in the Congress, in the agencies, in various agency meetings, at the White House . . . the person who shows up with the first draft seemed to be the one who would lead the others there."

Dembling was clearly determined to make the most of that practice and establish that lead. He recalled working late every night through January and February and often through the night, as well as weekends, scouring previous General Accounting Office decisions to find a form of words for legislation covering an entirely new area that would satisfy the BoB. But showing the connections and inner working that ensured legislation would be taken on board as intended, Dembling did explain that he was part of the BoB's board reviewing and approving the draft legislation. Yet there seemed to be no apparent conflict of interest in drafting the legislation and then being a part of its process of approval. Having a draft bill ready as soon as the president approved Killian's suggested way forward in establishing NASA was prescient, but it did not head off all the potential problems likely to arise from interested departments prior to the draft legislation reaching Congress. Dembling noted that two less than entirely transparent additions were made to the bill in the very brief time that both the DoD and Department of State had to review his draft. Responding to a question on whether there was a real desire to establish NASA as a purely civilian organization, in the fiftieth anniversary interview, Dembling said:

> I am very pleased that the Congress placed the act under the general welfare clause of the Constitution, and it stated that the aeronautical and space activities shall be the responsibility of, and shall be directed by, a civilian agency. The wording says "except the activities peculiar to or primarily associated with the development of weapons systems, military operations or the defense of the US shall be the responsibility and directed by DoD." DoD fought hard getting that provision into the bill. The Department of Defense said it would not cooperate unless it had that provision in the bill.[74]

There is no evidence that Eisenhower was troubled by this explicit coupling of NASA and the DoD in the Act—in fact, it could be argued that the language used provided the level of separation but joint endeavor that he wished to foster in both the nation's public and political mind. There is an oblique reference to the DoD's role in the notes Bryce Harlow prepared for the president's meeting with Johnson on July 7 when he discussed how to overcome the differences between the House and Senate's bills, especially on the make-up of the proposed Space Council. The notes stated:

> There is real concern in Congress about protecting the proper role of the military in our space activities. There is no essential difference in the stated purposes of House and Senate sponsors regarding the military. However, the actual language differs, we have taken the position (concurred in by Defense) that we prefer the Senate language, Insofar as it relates to military participation, the Senate language properly protects the responsibilities of the Defense Department.[75]

According to Dembling though, he did express some reservation around wording that the State Department insisted on adding to the original draft of the bill:

> The original bill provided that the administration could engage in a program of international cooperation. The State Department opposed that language and insisted that it had had to be under the foreign policy guidance of the President. . . . When the President signed the legislation he took a reservation and said something along the line that it is clearly understood that the international activities of NASA will be

done under the foreign policy guidance of the President and the State Department.

Two words may seem irrelevant but they are further evidence of the way in which Eisenhower's mind worked. While he was happy to share the acclaim when matters of policy worked out well—and indeed, was generous in his praise for those who contributed to the success—he was a canny president, who liked to have avenues of deflection in case things went awry. Thus, State had to be seen to have a joint responsibility for any prospective NASA international activities.

Indeed, the idea of international cooperation which finally found its way into the bill as it passed into law had not come from the White House as Glen Wilson explained in his oral history given to NASA. He noted that both the Senate committee under Johnson and the House Select Committee on Astronautics and Space Exploration under House Leader John McCormack worked to stamp their authority on the administration's initial proposal. Wilson recalled that both the House and the Senate made significant amendments to what had been passed over to them by the White House. He said: "Numerous modifications were made from the original proposal, including broadening and clarifying the scope of the space agency, changing its name to Administration instead of Agency and giving greater authority to the Administrator (instead of Director)."[76] Yet, that neither sounds nor feels like significant change, and in no way altered the thrust of the legislation. The House bill, introduced only on May 20 had cleared the House by May 24. Similarly, the Senate reported out their bill on June 11 and it was passed on June 16. In terms of Johnson's key contribution, Wilson says: "The Senate version broadened and clarified the scope of the agency, established a powerful Space Policy Board with a staff, authorized a program of international cooperation and retained the Joint Committee on Aeronautics and Space."[77] This is supported by Harlow's memorandum of the meeting between Eisenhower and Johnson on July 7, although the Space Policy Board was less powerful than the Senate had intended and Harlow noted that the version proposed by the Senate could "deny Presidential flexibility."[78]

MR. SPACE

It is important to be clear on Johnson's role in the spring and early summer of 1958 in relation to the Space Act. After a quiet period when his focus was on domestic affairs, Johnson had resurfaced as "Mr. Space" during the Senate's perusal of the proposed space legislation at committee stage. On May 6, 1958, Johnson's new committee, the Senate Special Committee on Space and Astronautics convened in Room 457 of the Senate Office Building to consider the NASA bill. Unsurprisingly, Johnson chaired the committee, and was joined by Senators Green, McLellan, Anderson, Symington, Bridges, Wiley, Hickenlooper, Saltonstall, and Bricker. Hickenlooper was a Republican, but an arch conservative. However, he had coauthored Eisenhower's Atomic Energy Act of 1954, and was seen as loyal to the president's wishes. The other Republicans were Saltonstall, who had overseen the Crossroads nuclear tests for Truman in 1946;[79] Wiley, who had previously chaired the Foreign Relations Committee;[80] and Bricker, whose "Bricker Amendment" motion in Congress, in 1954, attempted to limit presidential power over treaty making, had incurred Eisenhower's distinct displeasure.[81] Outnumbered five to four by Democrats anyway, the committee looked unlikely to be wholly favorable to the administration's ideas. The senators were supported by a committee staff personally selected by Johnson. Glen Wilson and Eilene Galloway had worked on the Armed Services Preparedness Subcommittee hearings; Gerald Siegel, de facto staff director, had been special counsel to the Democratic Policy Committee, and Solis Horwitz, was another of LBJ's Washington lawyers. The House committee had already reported out on the administration's bill, and the senators had the Dembling/NACA version and the House's amendments to work with in drafting their own version of the legislation. The committee conducted six days of hearings, interviewing twenty witnesses and compiling a two-volume, 413-page collected testimony.[82] Johnson, as ever, was visible and voluble. He began by saying:

> I believe it is entirely fair to say that seldom, if ever, has a Congress and an administration faced a more challenging task. We are dealing with a dimension, not a force. We are dealing with the unknown—not the known. While the present is urgent, the real imperative is the future. What we do now may very well decide, in a large sense, what our Nation is to be 20 years and 50 years and 100 years from now—and of no lesser importance, our decisions today can have the greatest influence upon whether the world moves towards a millennium of peace or plunges recklessly towards Armageddon. . . . The challenge of the space age, at the beginning now, is to open a new frontier and permits its use for peace.[83]

Johnson's somewhat polemical opening suggested he wished to give the impression that the future of the United States' journey into space lay in his hands and those of the assembled senators around him. Yet the actual debate through the hearings was far more concerned with the operation of the new agency as defined by Dembling, under the auspices of PSAC. As such, they actually gave Johnson considerable scope to demonstrate his bi-partisan credentials as, indeed, he chose to do in his opening statement.

> It is appropriate and heartening, I think, that we begin this work now on a base of unity and broad agreement, rather than on a base of dis-

> agreement and contention. I see no reason why this spirit cannot be maintained. The primary legislation before the Committee is legislation drafted by the advisers to the Chief Executive . . . I know, on the part of the sponsors and I believe on the part of the authors, there is full expectation that public examination and discussion of the terms of the legislation can contribute many strengthening recommendations . . . this committee wishes to confine its deliberations to the issues which are most pertinent . . . more than six months ago a committee of this Senate undertook an extensive and exhaustive study. . . . No substantial purpose would be served by devoting further time to repetition of such testimony. Furthermore . . . the House . . . is holding hearings in this same field. We are not here to duplicate those hearings but to act in accordance with the facts that are presented to us. . . . What is before us now is not a question of whether we should begin the orderly exploration of space but, rather, the question of how such exploration may best be directed and initiated.[84]

Here was Johnson not so much working for the president (although he was, in effect, carrying out his wishes), but for the good of the country. He could be *seen* to be the Master of the Senate, magnanimous in his management of the "broad agreement," but in reality, he had no intention of shifting gears and not supporting the president—and nor did he have any position of strength from which to oppose the executive-driven legislative thrust. Purcell's March document had created a strategy and organization that was difficult to argue with. The hearings gained a certain sense of urgency on their last day, May 15, when the Soviets announced their successful launch of Sputnik 3, weighing in at 2,900 lbs and nearly twelve feet in length.[85] However, there was no repeat of the shock of Sputnik 1, and the committee rapidly returned to its assigned task. While Johnson could, and did, apply the checks and balances of senatorial oversight, his committee and the subsequent Senate Bill really only challenged Eisenhower and PSAC in one area—with the Space Policy Board—soon to be renamed as the Space Council.

Eilene Galloway had been a National Defense Analyst, and during this time had published a report called *Guided Missiles in Foreign Countries* for both the House and Senate Armed Forces Committees.[86] This had brought her to Johnson's attention. Three days after Sputnik 1 launched, Senator Johnson telephoned saying: "Eilene, I want to make me a record in outer space and I want you to help me."[87] Galloway's first contribution was in naming NASA—and was made not through her work with Johnson, but when John McCormack, the leader of the House called her asking for advice on whether he should set up a House committee to look at the legislation. Galloway urged yes, and also urged him to change the "A" in NASA from "Agency" to "Administration." Without demur, McCormack instructed an aide to make the change which carried over into the Senate and the final Act.

During Galloway's attendance at the House hearings, which included long and detailed testimony from NACA chairman Doolittle and his deputy, the NACA director Hugh Dryden, she became convinced that the proposed seventeen-strong internal advisory board, built on NACA's previous structure would be insufficient to have "any clout at all over other agencies in Washington."[88] She communicated this to Johnson who was looking for ways to strengthen the bill, and, one might surmise, create an ongoing role for him related to space. While it was clear that the new NASA was set to emerge as a civilian agency, Galloway was concerned that what was needed was a top-level policy coordinating board for the total United States space program, both military and civilian. She noted that she had no "clout with McCormack with regard to the Space Council," it was "fertile soil" in the Senate because of their emphasis on national defense. As already stated, the initial idea for the Space Council did not find favor with Eisenhower. Mieczkowski wrote that he: "feared that it would add more bureaucracy, slow down policy formulation and wrongfully assume powers that should remain with the President."[89] Clearly, Mieczkowski is drawing directly on Harlow's memorandum of the July 7 Eisenhower/Johnson meeting here, and this almost directly repeats Harlow's notes for the president.[90] As ever, Eisenhower had no wish to confront Johnson with views that might cause an argument. Killian was dispatched to express the president's opinion, but Johnson was adamant that a Space Council, modelled on the NSC, was both necessary and a cornerstone to any legislation that might be passed. The president and the Senate majority leader settled their differences on Sunday, July 7, at the White House. The new Space Council would be chaired by the president and would comprise the NASA administrator, the AEC chair, the secretaries of State and Defense and four further members. Yet Eisenhower hardly convened the Council during the rest of his presidential term and recommended its abolition as he left office. However, the July 7 compromise ensured both of the big political beasts of Washington had achieved what they wanted on the road to the United States' first space legislation and, indeed, organization. Eisenhower signed the Act into law as Public Law 85–568 on July 29, 1958.

It is striking how similar the final Act sounded in comparison to Purcell's four reasons why the United States should explore outer space. While congressional intervention had amplified the original four purposes into eight clauses, their meaning had not changed greatly. The Act first stated:

> The Congress hereby declares that it is the policy of the United States that activities in space should be devoted to peaceful purposes for the benefit of all mankind.

It went on to say that:

The aeronautical and space activities of the United States shall be conducted so as to contribute materially to one or more of the following objectives:

(1) The expansion of human knowledge of phenomena in the atmosphere and space;
(2) The improvement of the usefulness, performance, speed, safety, and efficiency of aeronautical and space vehicles;
(3) The development and operation of vehicles capable of carrying instruments, equipment, supplies and living organisms through space;
(4) The establishment of long-range studies of the potential benefits to be gained from, the opportunities for, and the problems involved in the utilization of aeronautical and space activities for peaceful and scientific purposes.
(5) The preservation of the role of the United States as a leader in aeronautical and space science and technology and in the application thereof to the conduct of peaceful activities within and outside the atmosphere.
(6) The making available to agencies directly concerned with national defenses of discoveries that have military value or significance, and the furnishing by such agencies, to the civilian agency established to direct and control non-military aeronautical and space activities, of information as to discoveries which have value or significance to that agency;
(7) Cooperation by the United States with other nations and groups of nations in work done pursuant to this Act and in the peaceful application of the results, thereof; and
(8) The most effective utilization of the scientific and engineering resources of the United States, with close cooperation among all interested agencies of the United States in order to avoid unnecessary duplication of effort, facilities, and equipment.[91]

CONCLUSION

It is important to reflect on the rather more active role that Eisenhower played not in responding to Sputnik, but in charting a course for the US space program beyond the end of the IGY. Eisenhower's response was not perfect, but nor was it a passive passenger ride. As we have seen, Eisenhower was pursuing a parallel path in defining the civilian journey for the United States into space, closely aligned to a series of military missile programs that were gaining considerable success by 1958. While he had a genuine interest in scientific discovery, his primary driver was national security, but security gained and maintained by balancing resources against the reality of the Soviet threat. He had no interest whatsoever in weaponizing space nor in engaging in a fatuous race that seemed set only to define outer space as a proxy Cold War battlefield. In facing down the military establishment and heading off the Gaither apocalysts, he showed true leadership. In turning to the scientists of PSAC, and especially Jim Killian, he displayed the Eisenhower flair for masterful logistics: harnessing scientific goodwill, managed by a non-political, respected administrator to deliver a legislative outcome that promised to deliver a rational scientific space program bounded by sensible budget restrictions. Throughout the period, Eisenhower maintained the intelligence lead the United States had over the Soviets by not revealing his hand in terms of the knowledge he had of the fact that much of Khrushchev's sabre rattling was bluff. The Soviet leader had an effective heavy-duty booster, but nothing else technologically that was going to challenge the US militarily or in any other aspect of its hegemony.

He had recovered much of the ground lost due to the media-misstep by November, but was then hit by his stroke. Coinciding with the beginning of Johnson's Preparedness Subcommittee hearings, this could have been fatal in policy terms. However, the decision to empower Killian at the start of November with the task of charting the necessary policy and organization for scientific space exploration ensured that Eisenhower's illness did not impinge on his goals. What emerged, through the likes of "Helping Hands" Killian, Purcell, and Dembling, aided, not always consciously, by Johnson and his acolytes, was a direction for involvement in space and an organization to deliver it, that probably surpassed Eisenhower's aims for it as the thought of what to do post-IGY began to coalesce in his mind. Eisenhower was not passive. He may have been slightly wounded by the post-Sputnik hubbub, but he emerged stronger and with a very clearly-defined structure for progress.

One must understand too that this structure for progress was not set to deliver anything that was not anticipated as early as 1955 when the TCP reported. Eisenhower was able to ignore the hysteria created by other people's agenda setting. U-2 overflights continued to operate on his express orders; Bissell's CORONA team was progressing the development of the first spy satellites; rocket development was on track and the United States had its first satellites in space. None of these were catalyzed by Sputnik.

NOTES

1. R Melanson and D Mayers, *Re-evaluating Eisenhower: American Foreign Policy in the 1950s* (Urbana, IL, 1987), p. 1.

2. "Inquiry into Satellite and Missile Programs, Hearings before the Preparedness Investigating Sub-Committee of the Senate Committee on Armed Services, 85th Congress 1st and 2nd Sessions" (Government Printing Office, Washington, DC, 1958) p. 4. The sessions began on November 25, 1957, and ran until July 24, 1958. In total, the committee hearings produced almost 2,500 pages of testimony.

3. TM Gaskin, "Senator Lyndon B. Johnson, the Eisenhower Administration and U.S. Foreign Policy, 1957–60" in *Presidential Studies Quarterly*, Vol. 24, No. 2, Eisenhower and Governance (Spring 1994), pp. 348–49.
4. *Time* editorial, "Lyndon Johnson Moves in on Administration," *Time*, Monday, Jan. 20, 1958, (New York), p. 20.
5. R Dallek, *Lyndon B Johnson—Portrait of a President* (London, 2005), pp. 108–11.
6. Gaskin, *Presidential Studies Quarterly*, pp. 341–45.
7. RA Caro, *Lyndon Johnson Master of the Senate* (New York, 2003), p. 1021.
8. Ibid., p. 1024. Caro's ordering is not quite right. Von Braun did not appear in the first week of hearings. His voice was not raised as either an echo or amplification of Teller. Von Braun did not appear before the subcommittee until mid-December—and his alliance was not so much to the scientific as to the military establishment.
9. S.3609, 85th Congress. Congressional Record Y4.Sp1:Ae8/p.1 and 2, RG 287, US Centre for Legislative Archives, National Archives, Washington, DC.
10. *Legislative Origins of the National Aeronautics and Space act of 1958, Proceedings of Oral History Workshop*, moderated by JM Logsdon and published as *Monographs in Aerospace History Number 8* (NASA History Office, Washington, DC, 1998), Appendix A, p51.
11. Thomas, *Ike's Bluff*, p. 276.
12. S.36069, 85th Congress.
13. R Drummond, "Almost Not Good Enough," opinion column, *New York Herald Tribune*, syndicated from Washington, DC, November 28, 1957.
14. S.36069, 85th Congress.
15. L Thomas and E Jablonski, *Doolittle a Biography* (New York, 1976).
16. Doolittle joined the main committee of the NACA in the mid-1950s, following a spell as chair of the Air Force's Scientific Advisory Board. He became the NACA's Chairman in 1956—sourced from JH Doolittle with CV Glines, *I Could Never Be so Lucky Again* (New York, 1991).
17. S.36069, 85th Congress.
18. McElroy had been president of Procter & Gamble prior to his appointment to succeed Charles Wilson as Secretary for Defense. His career had largely been in P&G's marketing operations and his 30 year service to the company had been augmented only in 1954–1955 when he chaired the White House Conference on Education. Biographical information sourced from http://www.eisenhower.archives.gov/research/finding_aids/pdf/McElroy_Neil_Papers.pdf. Accessed March 20, 2013.
19. S.36069, 85th Congress.
20. Ibid.
21. Ibid.
22. Ibid., Gavin testified on December 13, 1957.
23. Ibid., Clarence testified on December 17, 1957.
24. Ibid., Medaris testified on December 17, 1957.
25. MJ Neufeld, *Von Braun* (New York, 2007), p. 317.
26. S.36069, 85th Congress.
27. DDE speaking at Graduation Exercises at the United States Military Academy, West Point, March 3, 1947.
28. S.36069, 85th Congress.
29. *Washington Post* Section 1, p. 1, December 15, 1957.
30. Neufeld, *Von Braun*, p. 317.
31. S.36069, 85th Congress. Putt testified on December 17, 1957.
32. DA Ritchie, ed., *Minutes of the Senate Democratic Conference 1903–1964*, S. Doc. 105–20, U.S. Government Printing Office (Washington, DC, 1998), p. 505.
33. S.36069, 85th Congress. Rockefeller testified before the Committee on January 10, 1958.
34. American Rocket Society, "National Space Establishment," January 4, 1958.
35. Divine, *Sputnik Challenge*, p. 67.

36. Sourced from Gallup presidential approval ratings at http://www.gallup.com/poll/116677/presidential-approval-ratings-gallup-historical-statistics-trends.aspx. Accessed, March 13, 2014.
37. *Time*, Volume 71, issue 1, January 6, 1958, p. 1.
38. As shown in United—international's newsreel footage on *A Time to Remember: The Space Race*, (AG Plate DVD, 2007) ch. 5.
39. *Daily Express* (London), December 7, 1957, p. 1.
40. *Daily Herald* (London), December 7, 1957, p. 1.
41. *Paris-Journal* (Paris), December 7, 1957, p. 2.
42. *Louisville Courier-Journal* (Louisville, KY), December 7, 1957, Section 1, p 2.
43. Question raised by Soviet Ambassador during UN General Assembly discussion of Resolution A/RES/1215(XII) on the United Nations Technical Assistance programs, New York, 14 Dec. 1957.
44. A Whitman diary entry for January 31, 1958, Box 9, Diary Series, Ann Whitman File, DDE Presidential Files, DDE Library.
45. Ibid.
46. The mandate to develop all land-launched missiles with a range of more than 200 miles had been given to the Air Force in 1955 as a result of the TCP recommendations. The Jupiter was, of course, an army missile developed at the Redstone Arsenal in Huntsville, Alabama, by Von Braun's team.
47. LA Minnich, "Legislative Leadership Meeting, Supplementary Notes," February 4, 1958, DDE Papers as President, DDE Files, DDE Library.
48. SP Johnston, "Memorandum for Dr. J. R. Killian, Jr.," *Activities*, February 21, 1958, with attached: Memorandum for Dr. J. R. Killian, Jr., "Preliminary Observations on the Organization for the Exploitation of Outer Space," February 21, 1958, Killian Files, MIT.
49. Eisenhower met with the GOP Leadership on February 4, 1958, and expressed his preference for DoD control over any future space agency. Box 9, Ann Whitman File, DDE Files, DDE Library.
50. LA Minnich, Jr., "Legislative Leadership Meeting, Supplementary Notes," February 4, 1958, DDE Papers as President, DDE Library.
51. AJ Goodpaster, "Memorandum of Conference with the President," February 4, 1958. DDE-EPRES, DDE Papers as President of the United States 1953–1961, DDE Library.
52. AJ Goodpaster, "Memorandum of Conference with the President," March 5, 1958. DDE Papers as President of the United States 1953–1961, DDE Library.
53. Department of Defense, DoD Directive 5105.15: Department of Defense Advanced Research Agency (Washington, DC), Feb. 7, 1958.
54. Goodpaster, "Conference with the President," March 5, 1958.
55. Ibid.
56. Kalic, *Presidents and Militarization of Space*, p. 3.
57. Vanguard budget information drawn from *United States IGY Bibliography, 1953–1960: An Annotated Bibliography*, Library of Congress, Science and Technology Division, (Washington, DC, 1963).
58. "Introduction to Outer Space, an Explanatory Statement Provided by the President's Science Advisory Committee," (Washington, DC, March 28, 1958). Sourced from the Federation of American Scientists' historical publications http://www.fas.org/spp/guide/usa/intro1958.html. Accessed on February 17, 2013, cross referenced with the same text reproduced in Killian, Appendix 4, pp. 288–99.
59. PSAC, "Introduction to Outer Space," pp. 11–13.
60. Ibid., p. 12.
61. Ibid.
62. AJ Goodpaster, "Notes on Outer Space Briefing . . . at Cabinet Meeting," Friday, March 14, 1958, DDE Papers as President of the United States 1953–1961, DDE Library.
63. Killian, *Sputnik*, p. 123.
64. Ibid., p. 124.

65. Ibid., p. 288.

66. Wernher Von Braun, the inventor of the V2 rocket had, by 1958, been in the United States working for the Army for twelve years and was a US citizen. In his early years in America he had grown an increasing media profile. At first, he lobbied for more funding and status for missile—and by implication space rocket—research. An opportunity to work with Walt Disney on national television network ABC did much to raise his profile while cementing the potential of spaceflight as an aspiration for America. In a bid to reinforce the promotion of his new theme park, Walt Disney linked up with ABC to launch a new TV show—*Disneyland*—in 1954. The format varied each week, ranging from animations to science instruction programs. By 1955 it was ranked by TV pollsters Nielsen as the fourth most popular show across the nation. Disney had read Von Braun's series of articles in *Collier's* Magazine, where the rocket engineer had sought to spread the space exploration gospel. The articles prompted two spin-off books, *Across the Space Frontier* in 1952 and *Conquest of the Moon* a year later. With his quirky German accent, opinion on everything and mission to explain, Von Braun became an obvious choice for Disney. Disney was not being altruistic: he was looking for a scientist to ally with his soon-to-be-opened "Tomorrowland" at his California theme park. Von Braun opened with a documentary *Man in Space*, following with *Man and the Moon* and *Mars and Beyond*. All three programs received excellent viewing figures in the United States and were significant in building the momentum to educate Americans in the wonders of the coming space age.

67. Janis, *Groupthink*, p. 37–39.

68. Killian, *Sputnik*, p. 132.

69. NA Rockefeller, "US Scientific Satellite Program," Memorandum for Mr. James Lay, Executive Secretary, National Security Council, May 17, 1955, DDE Files as President, 1953–61, DDE Library.

70. James R. Killian, Jr., Special Assistant for Science and Technology; Percival Brundage, Director, Bureau of the Budget; and Nelson A. Rockefeller, Chairman, President's Advisory Committee on Government Organization, "Organization for Civil Space Programs," Memorandum for the President, March 5, 1958, with attached: "Summary of Advantages and Disadvantages of Alternative Organizational Arrangements."

71. Killian, *Sputnik*, p. 134.

72. *Tucson Daily Citizen*, Friday, May 09, 1958, p. 9.

73. PG Dembling, quoted in E Glodstein, GC La Rosa and DS Schuman, "Present at the Creation," *NASA 50th Anniversary Magazine* (Washington, DC 2008), nasa.gov/50th/50th_magazine/dembling/interview.html. Accessed 21 February, 2013.

74. Ibid.

75. B Harlow, "Main Problems in the Senate Bill Establishing a Federal Space Agency," July 7, 1958, Box 9, Ann Whitman File, DDE Diary Series, DDE Library.

76. NASA, *Monographs in Aerospace History Number 8*, Wilson Appendix A.

77. GP Wilson, "How the US Space Act Came to Be," presentation paper from the symposium "The Legislative Origins of the US Space Program," held at George Washington University, April 3, 1992, p. 53.

78. Harlow, "Establishing a Federal Space Agency," July 7, 1958.

79. Massachusetts Historical Society, http://www.masshist.org/features/saltonstall/senator-saltonstall-bikini-atoll. Accessed September 19, 2014.

80. Biographical Directory of the United States Congress, http://bioguide.congress.gov/scripts/biodisplay.pl?index=W000465. Accessed September 19, 2014.

81. JE Smith, *War and Peace*, pp. 599–604.

82. GP Wilson, *How the US Space Act Came to Be*.

83. "National Aeronautics and Space Act, Hearings Before the Special Committee on Space and Astronautics, United States Senate," Eighty-fifth Congress, Second Session, on S. 3069, a Bill to Provide for Research into Problems of Flight Within and Outside the Earth's Atmosphere, and for Other Purposes. May 6, 7, and 8, 1958, Parts 1–2, Government Printing Office, (Washington, DC, 1958).

84. Ibid.
85. Hall and Shayler, *The Rocket Men*, p. 67.
86. Drawn from Eilene Galloways's individual discussion with John Logsdon at the symposium "The Legislative Origins of the US Space Program."
87. Ibid., p. 31.
88. Ibid., p. 33.
89. Mieczkowski, *Sputnik Moment*, p. 172.
90. Harlow, *Establishing a Federal Space Agency*.
91. "National Aeronautics and Space Act of 1958," Public Law #85–568, 72 Stat., 426. Signed by the President on July 29, 1958, Record Group 255, National Archives and Records Administration, Washington, DC.

FIVE

From Strategy to Implementation

Having a strategy is one thing, making it happen, quite another. It is time now to examine the outcomes enabled by the 1958 Space Act for military missile development and civilian space exploration. This will enable readers to understand the importance of the space legacy left by Eisenhower. Each space policy issue analyzed is important in itself, but there are wider considerations too. Using space policy as a case study contributes to our overall understanding of the late-period Eisenhower presidency and directly challenges the "complacency" described by Brogan and others.

Relatively little has been written about Eisenhower's interventions on space policy after NASA became operational, and there still remains a sense that by the summer of 1958, Eisenhower was a "lame-duck" President coasting toward retirement.[1] It is time to re-examine that interpretation looking at several key issues where Eisenhower did intervene in the application of his defined policy direction to ensure his national security-led and fiscally-conservative policy remained balanced. Equally, in reading the early space canon, it is easy to believe that there was no executive action on space until Kennedy addressed Congress in May 1961 and made his pledge to send a man to the moon. Again, this is simply not true. What *is* true is that space policy was not of the foremost importance to Eisenhower over the final two and a half years of his presidency. He was, however, determined to leave a legacy that reflected his presidential principles: moderate, conservative, balanced, and focused on waging peace. Actions including his appointment of NASA's first administrator and key decisions both to civilianize the development of the first US super booster and to assign the man in space program to NASA all demonstrate his ongoing strategic imperative to separate the civilian and military programs and also keep control of the implementation of the administration's new space policy. Meanwhile they also reflect his ability to make tactical changes within his overall strategy that demonstrated not an inflexible dogmatist, but true executive leadership, ultimately accountable for the strategic direction and key decisions, but equally willing to back the evidenced opinion of his expert "Helping Hands" in finding the best ways to implement that strategy.

TAKING CHARGE

Once NASA was established, Eisenhower's focus in terms of space policy was to ensure he had the right operational officers in place to deliver his strategy and the right operating structures to ensure both the scientific civilian program and the reconnaissance-led national security program could proceed effectively. Eisenhower's operational team, at cabinet and agency level had remained largely intact from when he entered the White House in 1953 until 1957. However, his second term saw a number of significant changes. The resignation and subsequent death of Secretary of State John Foster Dulles on May 24, 1959, was the most significant in terms of national security policy, and his death was a personal blow to Eisenhower. In a statement issued from his Gettysburg home on Dulles' death, Eisenhower said:

> [We] grieve at the passing of one of the truly great men of our time. Throughout his life, and particularly through his eventful six years as Secretary of State, his courage, his wisdom and his friendly understanding were devoted to bettering relations among nations. He was a foe only to tyranny. . . . We have lost a dear and close friend.[2]

Dulles was the latest in a succession of key personnel to leave the White House in the second term, following Herb Brownell, who retired as attorney general after seeing the 1957 Civil Rights Act onto the statutes; Sherman Adams who had been forced to resign over the vicuna coat affair and George Humphrey, the Treasury secretary who had resigned from office, also in 1957. Each had been a member of Eisenhower's White House inner circle of political influencers, either, as in the case of Humphrey, like-minded men with exceptional capability in turning Eisenhower's policy ideals into workable solutions, or, like Dulles, useful ideologues, capable of sparking strategic thought from Eisenhower or, on occasion, of being useful front men when the president needed to express strong opinions across the Iron Curtain.[3] With a gradually shrinking inner circle, Eisenhower turned ever more to those advisors beyond the political realm that he trusted most. PSAC was one group that, through the work of the TCP and through the scoping of NASA and subsequent legislative drafting, had proved its worth and earned the trust of the president.

Keith Glennan, who became NASA's first administrator, was a PSAC recommendation. He came out of the mold already inhabited by Jim Killian and approved by Eisenhower. He was a college administrator with Washington governmental experience and had served as an Atomic Energy commissioner. Yet he was not the first choice to be NASA's first administrator. The initial preferred candidate was General Jimmy Doolittle. On July 17, 1958, Killian met the president to inform him that the Space Bill had "passed the Congress in what seemed to be an acceptable form."[4] They then discussed who the first "Director" (Administrator) of the new NASA should be, and Killian suggested Doolittle. The memorandum, drafted by Goodpaster, noted: "The President strongly agreed, commenting that General Doolittle has the kind of force and energy required. If he is not available, Dr. Dryden, now the head of NACA, should be considered."[5] It is striking that Eisenhower continued the conversation on the clear assumption that Doolittle would take on the role and the president, showing military keenness to "develop an initial plan of operations," expected Doolittle "participating in its preparation." This made sense. Doolittle was the chairman of the NACA, a decorated war hero, who had been commanding general of the 12th Air Force in North Africa during the Eisenhower-led Operation Torch in 1942, and had followed him across the Mediterranean, taking command of the 15th Air Force. From January 1944 to September 1945 he commanded the 8th Air Force in Europe and the Pacific. He was one of Eisenhower's key operational commanders, and with a post-war career as a business executive with Shell Oil, he appeared to be a natural fit for Eisenhower's preferred way of working, in so far as he was used to the military chain of command both as a commander and an operational officer, but had added significant business management skills during his time at Shell.[6]

Yet Doolittle's time at the NACA had not been without controversy. Under Truman, he had been appointed a special assistant to the air force chief of staff, serving as a civilian adviser in scientific matters which led to air force ballistic missile and space programs. After Korea, where he returned to active service, he retained his Air Force rank and was perceived at the NACA to be partisan in favor of the Air Force over the other services, while also pushing the agenda of major business interests (not least through his Shell connection). This did not sit well in an organization that had, by tradition, been led by academics and had fought particularly hard to remain independent and above bias. According to NASA's own history of the NACA: "If Doolittle was anything, he was an academic last; first or second he was a businessman, second or first a military officer. He was the personification of what Eisenhower was soon to label the military industrial complex."[7] However, one could suggest equal negative bias on the behalf of the NACA. For the "pure" academics, the appointment of someone with commercial skills and a military drive was undoubtedly a threat to the clubbable, institutional atmosphere at the

NACA's Langley headquarters. The NACA/NASA assessment of Doolittle underestimates him academically too since he held a PhD in Physics from MIT.[8]

Certainly, at this stage, Eisenhower did not see any risk in appointing Doolittle, despite the likely conflict of interest in a staunch Air Force supporter taking over a civilian agency. Indeed, this serves to point up one of the president's blind spots. To him, Doolittle was a seasoned and proven military subordinate, one who had already proved himself as a loyal and capable leader tactically implementing the strategy put in place by the then supreme commander. In expecting and anticipating the same kind of relationship more than a decade later, Eisenhower was showing the same kind of cold logic that he brought to most decision making, while entirely failing to take account of the interservice political strife that such an appointment would bring. While Hugh Dryden, who had been director of the NACA was the second named potential NASA administrator, he was never a serious candidate. In his memoir, Killian noted that Dryden was generally considered too low key, and that in an atmosphere where Congress felt "space would revolutionize everything," he "did not have the exceptional qualities of enthusiastic leadership and drive" necessary to lead the space program.[9]

Meanwhile, on the role of NASA and what should be transferred out of the Department of Defense to the new agency, Eisenhower was unequivocal: "The President thought that anything not yet proved to technical feasibility should be the concern of this agency, and that non-military applications should also be the concern of this agency."[10] Essentially, the picture he had in his mind for NASA in mid-1958 was the first option for space and astronautical R&D, and the home of all civilian projects. Following the conference, Killian's role was to meet with Doolittle and encourage him to accept the NASA administrator's role. Killian is coy on the rebuttal he got from Doolittle. At sixty years old, one might have expected him to jump at this chance to take on an office reporting directly to the president. Killian remarked: "I later reached the conclusion that Doolittle should not be pressed because of his sincere reluctance to be considered." Doolittle's attachment to the Air Force may have swayed his refusal to take on the role. But it is more likely that he was swayed by the new senior role he had just taken on with Shell that was both demanding, and very well paid. His biographers offer no insight into why he chose not to move from the NACA to NASA.[11] There is also no explanation for his decision in the primary record. But Doolittle was not lost to the development of space policy entirely. He became one of the three "civilians" on the Space Council. In his diary, Glennan introduced him as "Lt. Gen [James H.] Doolittle, U.S.A.F. (ret)," so one might question just how "civilian" the civilian element of the council was! Hansen, in his history of Langley as the headquarters of the NACA and initially as NASA's major research base notes Doolittle's qualities and why Eisenhower was drawn to him for a position on the Space Council:

> Ike knew Doolittle . . . trusted his judgement; and wanted his moderate, reasonable and experienced voice on the newly formed space council.[12]

Scholars of early US space policy generally agree that Glennan was the right man to lead the kind of agency that Eisenhower had envisaged. Once Doolittle had turned the opportunity down, Glennan's name came to the fore in a number of conversations Killian had with his PSAC colleagues. The results of these suggestions were quickly relayed to the president. Even Roger Launius, usually critical of Eisenhower, recognized that Glennan had the attributes Eisenhower required.

> He was a Republican with a fiscally conservative bent, an aggressive businessman with a keen sense of public duty and an opposition to Government intrusions in to the lives of Americans, and an administrator and an educator with a rich appreciation of the roles of science and technology in an international setting.[13]

Glennan was an engineer with experience on the Atomic Energy Commission. Like Killian, he was highly regarded in academic circles as a capable administrator. As ever, the president wanted to judge the man in a face-to-face meeting before deciding if he could work with him. He had first met Glennan at an AEC meeting in 1955, but arranged a meeting through Killian for August 8, 1958. Glennan first knew of this the day before, and had little time to prepare for meeting the president. When Eisenhower sensed the right personal chemistry, his decision-making process was swift and easy and he offered Glennan the administrator role at the end of a meeting that Glennan recalls as: "very brief and very much to the point."[14] Two days later he accepted, and, on August 19, was sworn in at the White House, with Hugh Dryden as his deputy. Eisenhower's turn-around from looking to a decorated military leader to a career-bureaucrat to lead NASA reflected two things. First, his implicit trust in Killian's judgment. In his diary, Glennan noted that Eisenhower appeared to lean heavily on Killian's recommendations when they met.[15] Second, it reflected the style of agency Eisenhower expected NASA to be, and the direction it should take in space exploration. The program Eisenhower described to Glennan would be "sensibly paced" and Glennan noted that the president "made no mention of concern over accomplishments of the Soviet Union."[16] Glennan was not being brought in to lead any crash programs or speed ahead in any space race. Thus, capability and competence were the bywords for his appointment. On reflection, Eisenhower had once again taken the pragmatic route and recognized that there was no need for stars, particularly those with stars on their shoulders, for this particular appointment, although that pragmatism was made significantly easier by the fact Doolittle had rebuffed the offer in the first place.

In accepting the NASA role, Glennan had insisted that he would only become administrator if Dryden was to stay with the new NASA as his deputy. This proved an excellent decision bringing together Glennan's strength in working the Washington beat to build and shape the new organization, while Dryden, with his intimate knowledge both of the technical capabilities of the NACA and of the organization's people, brought a great strength in understanding the emerging science and engineering of space—and how best to channel and prioritize NASA's resources.

MAN IN SPACE

With Glennan in post, Eisenhower followed his previous operational practice, stepping back from day to day involvement with NASA and enabling his chosen operational officer to manage the agency within the president's mandate. Indeed, Glennan presided over a number of key NASA decisions, including defining the specific program for the agency and the necessary capabilities to execute it. Some decisions, such as the move of the Army Ballistic Missile Agency to NASA were long, drawn-out and contentious. But seemingly, one of the most important, the man-in-space program, was handed to NASA without resistance even before Glennan took up his new role.

In setting the remit for NASA as research and development relating to space and all non-military programs, Eisenhower had ensured that man-in-space was an obvious program for the new agency and, indeed, the NACA had been working on an outline for manned space activities even before NASA was mooted. As early as March, the Air Force had begun planning a "manned satellite development."[17] The NACA was acting as consultant to the Air Force, and proceeded jointly to develop the Air Force Manned Military Space System Development Plan.[18] This aimed to "achieve an early capability to land a man on the moon and return him safely to earth."[19] The plan was split into four phases. The first, "Man-in-Space-Soonest," focused on placing a ballistic capsule in orbit. The capsule would initially carry instruments, then apes, and finally a man. By the time Eisenhower had signed the Aeronautics and Space Act into law, he had already decided that the manned space program would be part of the civilian program, and thus the Air Force's work passed to NASA even before the agency became operational. Eisenhower's rationale was that it was a proof of concept program that would be run at minimal cost using, as far as possible, existing technology. After extensive discussions within ARPA and the Air Force's Air Research and Development Center, it had been decided that "Man-in-Space-Soonest" would use the existing Atlas booster as a launch vehicle. This played directly into that requirement for low-cost development and re-use of existing technology. Eisenhower's directive that NASA take on the program was notable for one slight change in wording that reflected both his cautious nature and lack of sentiment toward prestige achievements. He removed the word "Soonest" from the program.

Due to the extensive NACA involvement from March 1958 onwards, the transition of the program into NASA was smooth. Only five days after NASA was established, the Space Task Group was formed within it to design and implement the first US manned satellite. The STG was led by Robert Gilruth, an NACA veteran, and was to be based at the NACA's existing base, and the new NASA's primary research base, Langley, about three hours south of Glennan's headquarters in Washington, DC. The project was named "Mercury" since the symbolic associations of this name appealed to Abe Silverstein, NASA's director of space flight development, who suggested it in the fall of 1958.[20] Glennan made the name public on December 17, 1958. The smooth transition was one thing, dealing with the "manned" element turned out to be something quite different, and one of the most challenging aspects of NASA's early development.

However, in the fall of 1958, the new agency had other matters to deal with. In addition to Project Mercury, NASA had inherited ARPA's scientific earth satellite and lunar probe programs and the IGY Vanguard satellite program. Glennan's new staff amounted to eight thousand engineers and support staff, most of whom were highly skilled and highly qualified technicians. But the large majority had also been operating in a research backwater since the end of World War II. Even if Glennan was to meet Eisenhower's relatively modest and conservative goals for space, he needed some significantly increased engineering and scientific firepower. At that time virtually all of that was housed within the military. This would be a substantial test for Glennan and one that Eisenhower fully expected him to resolve. Eisenhower's management style demanded that his subordinates brought him solutions rather than problems, and he chose only to get involved when the problems were so serious that they demanded presidential-level action.

After a visit to the Redstone Arsenal at Huntsville Alabama, where Wernher Von Braun's ABMA team was based, Glennan became convinced that: "the nation's space program would advance most rapidly if we had working within our framework the so-called Von Braun team at Huntsville, and the Jet Propulsion Laboratory (JPL) operated for the Army by Caltech."[21] Once he had come to this conclusion, Glennan soon realized that he faced strong opposition from the Army, which was loathe to give up its key space assets, and also that he had scant support from his new NASA staff, whose focus had for so long been on consultancy, theoretical planning, and component testing. The former NACA had also long resisted getting into fights with any part of the DoD. Yet, with agreement from Killian and Eisenhower, Glennan took on the Army and soon saw at first hand the dysfunctionality of the DoD where one thing was said to the NASA's administrator's face, while quite another thing happened in practice.[22]

FOOLS RUSH IN

While the Mercury man-in-space program was bound to be the aspect of NASA's work that attracted the public's attention, it was never going to become more than a proof of concept if it had to rely on the very limited booster power of the existing Redstone and Jupiter rockets, and the soon-to-be-operational Atlas. Thus logic dictated that NASA should take control of the means to produce a booster system capable not just of breaking out of the earth's atmosphere, but out of the earth's orbit, if manned space flight was ever to reach the levels of interplanetary exploration as envisaged in Purcell and York's *Introduction to Outer Space.*[23] The political reorganization of the space landscape through the Aeronautics and Space Act and the 1958 Defense Reorganization Act had placed research for civilian projects in NASA's hands, while the militarized use of space was now firmly in the hands of the Air Force, at this time through its Advanced Research Projects Agency (ARPA). Squeezed in the middle was the Army and its principal spacecraft, rocket, and ballistic research and development *and* engineering and production facilities. This was the Jet Propulsion Laboratory, operated for the Army at the University of California Institute of Technology (Caltech), and Von Braun's Army Ballistic Missile Agency (ABMA) at the Redstone Arsenal, Huntsville, Alabama.

Moving these facilities from military to civilian control was always going to be an issue, not least because the prospective moves came on the back of the 1958 reorganization of the Department of Defense. Eisenhower's key legislative battle in the wake of Sputnik had not been the creation of NASA, but a battle to reorganize the defense establishment through "clear organization and central direction" that would deliver unity, efficiency, and unchallenged civilian authority.[24] Having put in place the New Look defense strategy, based on massive nuclear deterrent delivered initially through SAC, and increasingly through ballistic missiles, Eisenhower's next task was to make the machinery for delivering this strategy both more efficient and effective. That meant overhauling the Pentagon: reducing the powers of each individual service; cutting duplication; removing gaps between the services and integrating those services under a more powerful secretary of defense to ensure that US forces could meet the challenges of the Cold War. In his news conference of April 2, 1958, Eisenhower explained his rationale:

> In modern times there is no such thing as a separate ground, air, or sea war. The defense of the United States requires the planning for and if necessary the use of all of our defense forces as an integrated team. This places before a centralized authority, the Secretary of Defense, the task of making strategic plans. To conduct or to execute strategic plans, you have to have unified commands. We have those now. But those unified commands, if they are to be responsive completely to the decisions of the Secretary, must be organized by him, their missions must be given, he must determine their strength, the composition of the forces that will be capable of carrying on the defense of the areas or area that he may prescribe.[25]

This was the message he took to Congress the following day in the midst of a battle both with the services and with congressional politicians. The armed services argued that reorganization of the Pentagon was unnecessary and that Eisenhower was intruding in areas best left to the military (an irony, given that he was probably the most militarily-knowledgeable commander in chief in the history of the Union), while Congress felt that Eisenhower was maneuvering to bypass Congress' role in the defense of the nation by putting all military decision making (and primarily all decisions on appropriations) in the hands of the secretary of defense who reported not to Congress, but directly to the president.

In the wake of Sputnik, Eisenhower held the high ground on the defense capability argument. Modern warfare required speedy, integrated decision making. Under the congressional/Joint Chiefs of Staff system, that simply was not possible. Throughout the spring and early summer, Eisenhower galvanized support for modernization against such significant voices in Congress as Carl Vinson, chairman of the house committee on armed services and Richard Russell in the Senate. He met frequently with senior Pentagon and congressional figures throughout the legislative campaign and organized letter-writing campaigns to lobby Congress in support of the executive's views. Learning a lesson from his somewhat lack-luster immediate response to Sputnik, he was also proactive in making his case with the media. He spoke on April 17, 1958, at a joint luncheon for the American Society of Newspaper Editors and the International Press Institute and said:

> The Congress will keep, in every respect, its full constitutional authority over the appropriation of funds. But greater flexibility in defense spending will result in greater efficiency, more responsiveness to changing military requirements, and more economical management of major defense programs.
>
> Apprehensions such as these are at the least misconceptions. At the most they are misrepresentations. I repeat, there will be:
>
> -no single chief of staff;
> -no Prussian staff;
> -no czar;

> -no 40-billion dollar blank check;
> -no swallowing up of the traditional services;
> -no undermining of the constitutional powers of Congress.
>
> But there will be, if the program which I so earnestly support and believe in is adopted by the Congress: there will be a stop to unworthy and sometimes costly bickering. There will be clear-cut civilian responsibility, unified strategic planning and direction, and completely unified combat commands.
>
> There will be a stop to inefficiencies and needless duplications encouraged by present law.
>
> Thus we will meet our dual needs-safety and solvency. The Congress willing, we shall have maximum strength, with minimum cost, in our national defense.[26]

His letters too, provide a fascinating insight into the corporate view of politics Eisenhower favored. He argued that businesses operated on the basis of maximum efficiency and minimum cost, yet the Department of Defense, floundered in disarray. Eisenhower's pro-forma letters set out a picture whereby the Pentagon would be structured along the same lines as a major industrial organization, operating through a board of directors headed by a CEO who would lead the decision-making process within parameters agreed by the board. In this case, the secretary of defense would be that CEO. He concluded by saying:

> If this little comparison with corporate practices appeals to you as helpful in appreciating the crying need for defense modernization, I hope that you and others will find it useful in awakening the public to the grave seriousness of the matter.[27]

In McElroy, he had a ready-made CEO as he had been recruited to Washington from the presidency of Procter & Gamble. Eisenhower's policy of inducting expertise from the corporate world was also reflected in Glennan's planning for NASA. Indeed, on taking up his duties as administrator, one of his first actions was to send the management consultants of McKinsey & Co. into the NACA to look at its structure and working practices and to plan for its new operation as NASA.[28]

In driving through the Defense Reorganization Act, Eisenhower won the public relations battle. But the fight in Congress was tough and ended in compromise on both sides. Theoretically, the Act gave more power to the secretary of defense, but Congress retained almost all of its powers of oversight. Meanwhile, actually implementing the new law in the services themselves would prove extremely challenging. It is notable that while the final Senate vote supported the executive proposals by a margin of 80–0, there were sixteen abstentions. Among them was a certain John F Kennedy, Senator for Massachusetts.[29] Eisenhower's pro-active stance showed that he was far from a "lame-duck" even so late in his second term and faced with a hostile Democratic Congress. Unlike in his first term in office when he still had an election to win, Eisenhower could fight harder and press his case more strongly. Quite simply, he had nothing to lose.[30]

As in most of his major political battles, Eisenhower succeeded through the strength of his personality and his public appeal. Throughout the first half of 1958, he never wasted the chance to make his case for defense modernization, and took as many opportunities as possible to explain his case either to key individuals in one-to-one meetings or to major audiences such as the United Republican Dinner in the wake of the State of the Union.[31] However, when his appointee Glennan approached the Army with regards to the JPL and ABMA, he was tactically less astute than the president, and significantly less successful. The New Look policy had been a blow to the Army, reducing its numbers and denting its prestige. The Air Force, and particularly SAC, was responsible for nuclear weapon development, while also taking control of ballistic missile development and military initiatives in space. In handing over Vanguard to NASA, the navy had relinquished its space technology development mission, though it would be involved in missile development, specifically in preparing a launch platform for the Polaris missile. While the Jet Propulsion Laboratory (JPL) was not integral to the Army's mission, the ABMA was extremely high profile. With its Jupiter missile development and the launch vehicle for the Explorer satellite, all under its technical director, Von Braun, it was a successful operation that the Army was extremely loathe to lose. In one corner, Glennan faced negotiation with the Army which had lost autonomy within the Pentagon and authority to Eisenhower through the Defense Reorganization Act. In the other, his opponent was not Von Braun who was expedient in his response to authority and was only looking for a "rich uncle," but his military boss, the martinet general, John Medaris.[32] While Glennan's visit to Huntsville had convinced him that Von Braun's group would be useful to NASA and that with the JPL it would provide a framework to advance the nation's space program, he faced significant problems. Given that Eisenhower had decreed that NASA was to be a civilian agency, there was no way for Glennan to accommodate an Army-led rocket development team. Glennan described his own follow-up moves to prepare for the transfer of the JPL and Von Braun's team from the ABMA to NASA as "naive" and "stubborn" but he did note that he spoke with both Killian and Eisenhower and gained their support before he approached the Department of Defense.[33] In hindsight, he should have spoken more about the tactics to use to win over the secretary of the Army. No doubt Eisenhower and Killian were very good sounding boards, but they gave only tacit support, and left Glennan to blunder into confrontation with the secretary of the Army.

Simplistic accounts of the early space narrative tend to put the genesis of NASA at the highest national priority. But if one was to look at Eisenhower's priorities in late 1958, and also at the many tasks assigned to PSAC, one would see that the decision on what to do with the Army's space technology assets was just one of a large number of issues clamoring for attention. The Middle East and Berlin dominated the president's agenda, with the Cuban revolution and even the admittance of the new States of Hawaii and Alaska competing for consideration. Equally, he had to deal with Khrushchev's almost constant carping throughout 1958 for the United States and the United Kingdom to impose nuclear testing bans as the three nuclear powers inched toward a formal test ban treaty. Meanwhile, Kistiakowsky provided an insight into the PSAC agenda with his diary comment on the working groups established under its authority:

> PSAC had incorporated a number of working groups: antiballistic missiles, chaired by Wiesner, antisubmarine warfare under Harvey Brooks, arms limitation and control chaired by Killian, basic research and education under Glenn T. Seaborg, chemicals in food (Bronk), continental air defense (Emmanuel R. Piore), early warning (Wiesner), high-energy accelerator physics (Piore), life sciences (George W. Beadle), limited warfare (H.P. Robertson), missiles (Hendrik W. Bode), science and foreign relations (Bronk), space sciences (Edward M. Purcell), an ad-hoc study of missiles under Donald P. Ling, and a panel to review military communications under William O. Baker.[34]

NASA was competing in a crowded arena of scientific advance, and if Glennan thought he could have the two army facilities transferred into NASA in the first months of the new agency, he was much mistaken.

Problems had started with Glennan's first meeting with Wilber Brucker, the secretary to the Army. Glennan had followed the president's normal practice by meeting first with Secretary of Defense McElroy and his assistant, Quarles, to reach an informal agreement with them on the way forward and to gain their support. Both had appeared supportive of Glennan's aim of building the nation's space program on the Huntsville/Caltech team. But the distinction should be drawn between McElroy and Quarles who sat at the top of the Pentagon pyramid above all the services, and Brucker who was responsible solely for the Army. Glennan noted that Brucker became "irate" at the suggestion to move the ABMA to NASA.[35] Glennan remarked that he left that particular meeting "with my tail between my legs."[36] Brucker then took the initiative, leaking the story to the *Baltimore Sun* which ran a story the following day stating that NASA was intent on breaking up the Von Braun team.[37] Glennan had handed the Army the upper hand. The ABMA in total amounted to over five thousand people, and the first NASA budget, agreed by Congress and approved by Eisenhower, would enable fewer than half the manpower to transfer.[38] But Von Braun was a hero in the eyes of the public and Brucker's leak (the *Baltimore Sun* story was picked up by other newspapers and ran around the country) ensured NASA was cast as the villain of the piece in threatening to destroy the team that had finally put America into space.[39]

However, it would seem that Glennan learned fast. He enlisted the support of Killian and Quarles, and both were effective into corralling the Army into line—at least in part. By November 1958, the Army had agreed to give up the JPL to NASA, but was refusing to give up the ABMA. It did, however, enter into an agreement with NASA for the civilian agency to use its services. This was an incomplete and unsatisfactory solution.

The following nine months proved to be difficult, especially for Glennan. The relationship between the ABMA and NASA was a significant stumbling block for NASA in establishing itself as the sole agency tasked with managing the United States' civilian space program. Von Braun would align his support with whoever provided him with the money and remit to continue his work. But General Medaris used every possible means to retain his hold on missile and rocket development. Glennan certainly had a sense through most of 1959 that NASA's mission was in real jeopardy due to Medaris's intransigence. It is surprising that the wrangling between the Army and NASA does not receive more coverage in the current histories. Even in Mieczkowski's most recent work, it rates only a paragraph.[40] However, it is important in understanding both the role Eisenhower played, and his expectations of those who were carrying out his orders. As with his time as an army general, Eisenhower expected adherence to his orders, and for his operational officers to resolve any issues standing in the way of the execution of those orders. For the moment, the ABMA was Glennan's problem.

Glennan's writing alludes to the struggle he faced in dealing with Medaris. One of the issues faced (which actually turned into an opportunity) that might have compounded the problem was the sudden death of Deputy Secretary of Defense Don Quarles in May 1959. While the Secretary of Defense Neil McElroy was rather remote and disconnected from the feuding between the services, Quarles had been an informed ally of Glennan and it appeared his sudden death of a heart attack would be a blow. However, he was succeeded by Tom Gates, former secretary of the navy, who had no particular love for the army. While Brucker continued to promote the army's case for retaining the ABMA in Washington, Medaris pulled the forces of army propaganda together to state the case for the army's continuing role in long range missile and space rocket development. In 1956, the American public had been introduced to the ABMA team for the first time through *The Big Picture*, a half-hour prime time TV series which was aired on ABC television. While ABC was still a struggling cousin to better-resourced NBC and CBS, this suited the army since in return for a prime viewing slot it provided a weekly ready-made thirty-minute program at no cost to the broadcaster. Syndicated across its network, the ABC broadcast would still have reached a very significant number of the fifty-four million TV sets in use in 1958 in the United States.[41] Now, *The Big Picture* returned to Huntsville in 1958 in the wake of the successful Explorer 1 and Explorer 3 launches to present a very pro-Army view of the ABMA's role in delivering the launch in eighty-four days.[42] It provided excellent evidence for Medaris to use as he tried to ensure the Army's place in NASA and the DoD's plans for rocket and missile development.

However, Medaris was becoming isolated as support for the Army's key rocket project, Saturn, waned within the DoD. At this time, the development of the Saturn booster was being handled by the ABMA, but the funding was coming from the ARPA which was facing resistance to the whole project from the Air Force. The Air Force was moving ahead with its ICBM development and had no need for a million pound plus booster within its military space plans. Even the Army had only one planned use for the booster: to launch the communications satellite the Signals Corp was building for the DoD. There was a strong possibility that Saturn would be cancelled.[43] Indeed, at this stage there were three super booster systems competing for adoption, and Eisenhower appeared to favor the alternative solid rocket booster, Nova, which had been presented to him by NASA in January 1959.[44] Medaris still appeared to be fighting a rear-guard PR battle when Glennan returned to Huntsville in September 1959. By now, the NASA administrator had learned from his experience of dealing with the Army the previous year. He had talked to Gates and: "made it clear that I proposed to make a new deal [to secure the ABMA for NASA] only with the Office of the Secretary of Defense and that I expected Brucker to be told the results once it had been made."[45] That deal-making had been making progress, with robust intervention from Gates who essentially took over the lead role in dealing with the Army. However, it was clear that news had filtered to Brucker—and indeed made its way to both Medaris and Von Braun. Neufeld noted that once the news filtered that it was likely that Saturn would be cancelled if the Army retained the ABMA or, indeed if the ABMA was taken over by the Air Force, several of the major figures in Von Braun's team decided they would be better off allied to NASA. Glennan "had a real valid requirement for Saturn."[46] His purpose, on his September visit, was to inspect progress on the Saturn booster project and he still expected opposition from Medaris who acted, Glennan thought, "as if he had little knowledge of what was going on."[47] However, Medaris was fully aware of the political machinations of dividing the space program up between the Air Force, reporting to ARPA, and NASA, reporting direct to the president. Glennan recorded:

> During the course of our visit to several shops, I noticed that he was called away to the telephone on several occasions. Finally, he called me

> to one side and asked me to come to his office . . . Word had leaked that Saturn might be abandoned. Medaris revealed that he had some knowledge of the impending transfer and stated that if I would agree for Saturn to be retained, he would personally go to Washington and urge that transfer.[48]

Medaris had been backed into a corner, but it is notable how the lure of a legacy can lead even the most extreme opponents of a new organization to appear to turn 180 degrees in their views. But could Medaris have believed that his action would ultimately keep the ABMA as an Army asset, or was he simply acting petulantly in his delaying tactics? He may, indeed, have had a sense that the political tide had turned even further against Eisenhower—or particularly, the wider Republican movement, and especially those aiming for Congress.

The 1958 Congressional elections had further cemented Democrat control of Congress, with a fifteen seat gain in the Senate, taking their representation to sixty-four Senators against the Republicans' thirty-four. In the House of Representatives, the Republican demise was just as startling as the Democrats gained forty-nine seats to hold a 283–153 majority.[49] Medaris may well have felt emboldened to hold out against the executive in the belief that Eisenhower would yield to the congressional clamor for more spending on defense. If that was the case, and there is no significant primary source evidence to either back or challenge such an assertion, he was severely misguided.

A year after Glennan first attempted to move ABMA into NASA, Kistiakowsky, who had by then taken over from Killian as chair of PSAC wrote in his diary:

> A meeting with the President. . . . He flatly stated that ABMA should be put under the NASA, and on my warning conceded that he will have to defend Glennan publicly. He expressed highest confidence in Glennan and said he understood the pressures under which Glennan was operating.[50]

Eisenhower's policy priorities placed national security at the heart of his planning and he directly drove national security policy, even on its fringes where elements fell into NASA's authority. Throughout the 1950s, national security was executive-led, essentially through building on Roosevelt's and Truman's previous policies. The key relationship affecting the ABMA was not congressional, but the interaction of the four actors of the White House, PSAC, NASA, and Pentagon. While having Brucker's support enabled Medaris to delay the inevitable, a major general and the secretary of the Army could not succeed in challenging the will of the commander in chief. Eisenhower, once again, was prepared to play the long game. On 20 October 1959, Glennan wrote formally to him regarding the transfer of the operations division of ABMA, couching the note in a discussion on budgetary figures. He wrote that such a transfer would:

> tend to further concentrate responsibility for our program of scientific research and exploration in the civilian agency established in response to your request to the Congress in early 1958.
>
> . . . I have long been concerned about the "rate and scale" of our non-military space effort . . . I now hold firmly the opinion that the United States cannot withdraw from the contest for significant progress in scientific research in space and the exploration and exploitation of space for purposes beneficial to mankind. But I am equally certain that we must develop and adhere to our own vigorous program rather than compete on a shot-for-shot basis with the USSR.
>
> It seems clear additional levels of funding will be necessary. . . . This is occasioned by the fact that rocket booster systems necessary to carry out increasingly difficult experiments in space and to provide ultimately for manned exploration of outer space and the nearby celestial bodies are costly to develop. It will be our purpose, ultimately, to center in the group transferred from ABMA to NASA the bulk of the management responsibility for our effort in the space development field.[51]

The letter proceeded to detail the cost of the transfer, who would pay what, and reach a conclusion that despite the seeming increase in NASA's budget, transferring the core ABMA team, in effect the Saturn booster team, to NASA "could effect savings of perhaps as much as 75 millions by consolidation of our work effort as a result of the proposed transfer."[52] After attempting to resolve the ABMA situation through negotiations at ABMA level, with the secretary of the Army and at DoD level, Glennan had finally found the formula that would prompt Eisenhower to intervene on his behalf: a budget saving and a clear purpose for the Saturn booster within NASA. It probably took Glennan too long to get to this solution, but he had a wily opponent in Medaris.

Eisenhower's response shows mastery in keeping above the debate while enabling Glennan to achieve the president's aim. In a letter forwarded to the media to share with the public, he entirely avoided any notion of a contest and wrote:

> Dear Dr. Glennan: As we have agreed, it is essential to press forward vigorously to increase our capability in high-thrust space vehicles. You are hereby directed to make a study, to be completed at the earliest date practicable, of the possible need for additional funds for the balance of FY1960 and for FY1961 to accelerate the super booster program for which your agency was given technical and management responsibility.[53]

This is classic Eisenhower. He focused on the budget. He did not make a final and binding decision, but asked for a study that would supply evidence for the need for such a decision. He was asking for clear and tangible evidence to justify significant spending. Such evidence would be ammunition to use with the other actors involved in the ABMA/super booster discussions. But he also made it plain that NASA was responsible for the super booster program. The Army was not mentioned but the intent for NASA to manage the Saturn program was clear.

Eisenhower finally signed the executive order transferring the necessary elements of the ABMA to NASA once the ground had been cut completely from beneath Medaris's feet. Before so doing, he met with Von Braun on October 21, 1959, and assured him of his support for the Saturn program. The order, and the transfer of the ABMA team to NASA came into effect on March 15, 1960. Brucker remained as secretary of the Army until the end of Eisenhower's second term, but the president was no doubt always aware that Wilbur Brucker had never risen above the rank of first lieutenant while a serving officer in the Army.[54] This was not a battle of equals. Medaris retired at the start of 1960 as the bulk of the ABMA was transferred to NASA by executive order.

There was, for Eisenhower, a very apt conclusion to the ABMA transfer when, under Executive Order 10870, signed on March 15, 1960, he designated the facilities of the National Aeronautics and Space Administration at Huntsville, Alabama, as the "George C. Marshall Space Flight Center."[55] Marshall, who had died the previous October, had no impact on the US space effort, nor did he have any links to Huntsville. But he was an army man through and through, and it was appropriate for his memory to live on adjacent to the Redstone Arsenal. The Marshall Space Center duly opened for business on July 1, 1960, and Eisenhower was there to perform the official opening. In the process, he directly aligned himself with Saturn's designated mission: the human spaceflight program.

Glennan's experience with the Army during the first year of NASA's operation might have been eased if he had sought Eisenhower's intervention earlier. But after Eisenhower's acknowledgment in December 1958 that the ABMA could remain within the Army for a further year, the task facing Glennan was to find a solution to the Von Braun/Medaris issue, not to land another problem on Eisenhower's plate. That would have been an admission of failure. Without the benefit of hindsight, Glennan could not see at the end of 1958 that the Medaris vision for the ABMA was doomed. Eisenhower had stated conclusively that manned space exploration would be a civilian activity. Manned flight was the only reason a Saturn-type super booster was needed and the ABMA made sense as an organization only if it was the means to deliver that booster.

Up until October 1959, Eisenhower remained aloof from the bureaucratic tussle. But that is not to say he was disengaged from the process or inactive. Glennan noted that he saw Eisenhower in his office "every four weeks and at intervals at meetings of the cabinet and NSC."[56] He added pointedly that Eisenhower appeared particularly "on top of his job, more decisive when matters were brought to his attention," once Sherman

Adams had resigned from the White House.[57] However, Glennan gave a telling insight into Eisenhower's modus operandus in closing his thoughts on engagement with Eisenhower by saying: "I tended to work closely with his [Eisenhower's] staff and particularly with Stans [Budget], Persons [Adams' successor], Killian, and later George Kistiakowsky, his General Counsel, Gerald Morgan and certain members of the cabinet."[58] This was another reflection of the necessity of the supporting cast of operational officers to deliver the operational tactics directed by the president's strategy. Eisenhower had replaced his battlefield commanders with White House equivalents. Yet what matters is that he was at the top of this pyramid structure, always guiding and always in control. The Whitman files demonstrate that he met with Persons on a daily basis, often several times a day and had very regular face to face meetings with each of the other named advisers, including Glennan.[59] Eisenhower was constant in his management style. He intervened, but only when it was necessary so to do. In the case of the ABMA, he trusted Glennan to achieve as much as possible on his behalf up to the point where it was necessary for Eisenhower to take decisive action.

PRESIDENTIAL IMPACT ON MERCURY

The Mercury program was assigned to NASA in the first week of its existence, and NASA had initially planned an open recruitment process to find its first group of astronauts. However, this plan was abandoned when Eisenhower agreed with Glennan, Dryden, and the STG Director Gilruth, that the group should be drawn from the existing cadre of military test pilots.[60] This was not because of any particular affinity to service pilots on the part of the president. As with many of Eisenhower's decisions, it was more mundanely pragmatic and based, in part, on the successful experience of recruiting U-2 pilots for the CIA over-flights of the Soviet Union. A panel consisting of Stanley C. White, a senior medical practitioner in the US Air Force; Robert B. Voas, assistant for human factors, office of the director; NASA Manned Spacecraft Center; and William S. Augerson, a senior medical practitioner in the US Army representing NASA and the military personnel bureau in Washington screened 508 test pilot records in January 1959 and 110 men met the minimum standards for astronaut selection. The process that whittled these 110 down to the seven presented in Washington, three months later is the subject of almost every history of the early space age, but few recognize that the process was essentially the same as that used to select the first U-2 pilots. The direct link between the U-2 and Mercury selection processes was the Lovelace Clinic in Albuquerque, where both the U-2 recruits and prospective Mercury astronauts undertook a week of intense medical evaluations.[61] While Eisenhower's U-2 chief planner Dick Bissell had managed the development of the spy plane in conjunction with Lockheed under the auspices of the CIA. Curtis LeMay, head of the SAC, insisted the Air Force train the pilots, even though they were to fly as CIA and thus civilian operatives.[62] However, the air force tests were not sufficiently stringent for final selection, and as a consequence, the prospective U-2 pilots were referred to Dr. William "Randy" Lovelace, a World War II Army Air Corps colonel, who had continued to conduct research in private practice following his demobilization, helping to improve aviation and aerospace medicine. The test regime established by Lovelace was deemed successful by the CIA and ensured he was at the forefront of NASA's thinking in 1958 as the newly created space agency began to develop its man-in-space program. Indeed, Dryden asked him to chair its Special Advisory Committee on Life Sciences.[63] The next year, a group of Lovelace Clinic aviation medicine experts under Dr. Lovelace's direction put thirty-one astronaut candidates through a week of clinical tests, in Albuquerque. On April 9, 1959, the best responders, the original Mercury Seven, were introduced to the media. Lovelace was on hand and described the demanding testing undergone by the astronauts. The Astronauts described the Lovelace tests as the most rigorous and grueling part of the selection procedure.[64] The U-2 selection process had worked. While the helping hand mantle had passed from Bissell to Glennan, there was no need to invent new practices. The lessons from U-2 were learned and applied to Mercury.[65]

A CHANGE OF FOCUS: KILLIAN TO KISTIAKOWSKY

Eisenhower's ultimate "Helping Hand," Jim Killian, left his post as Eisenhower's special assistant for science and technology on July 15, 1959. He was succeeded in the post by George Kistiakowsky, a noted Harvard chemist and leading member of Killian's PSAC team. Killian returned to his previous role as president of MIT, a planned move that Eisenhower had known would happen and had understood since the beginning of Killian's PSAC tenure in November, 1957. The succession marked a change of pace and emphasis for PSAC, and reflected an equal change in the status of the civilian space program. Killian's great skill, which had attracted Eisenhower to him, was his ability to bring the right people together to achieve great results. He was not particularly creative, nor was he a scientist. But, like the president, he was a master logistician. On Eisenhower's behalf he had marshalled disparate strands of thought, disparate bursts of activity and the competing egos of a swathe of highly intelligent alpha males into concrete action on behalf of the president. First, he had put PSAC's resources behind the formulation of the National Defense Education legislation. The claim that the Soviets were producing more engineers and scientists than the Americans had been levelled long before Sputnik 1 started orbiting the earth. But the Soviet success sharpened the crisis of confidence in US education. Killian's significant success in feeding into the developing administrative program for science education was to urge a drive for quality over quantity. As the college president who had introduced a faculty of arts into MIT, it should be no surprise that he lobbied for investment in modern languages alongside science, technology, engineering, and mathematics.[66] His meetings with Eisenhower and the president's brother Milton, a close adviser and another college president,[67] helped cement Eisenhower's view that basic scientific research and education required greater funding (as opposed to applied technology for instance through the defense industry), and consequently introduced programs, passed by Congress into law, to increase funding to the National Science Foundation.[68] Killian was also instrumental in creating a panel tasked with reopening negotiations with the Soviets on limiting nuclear tests. Its success was in presenting technical advice on the possibilities of measuring testing activity, and this led to the relatively successful Geneva Conference in 1958. Yet this was an area where PSAC was in danger of being mired in politics. It worked most effectively when operating as an apolitical advice group. With the competing cohorts of the DoD, the Atomic Energy Commission, and those elements of industry supporting both monoliths, PSAC became less useful to Eisenhower following the Geneva Conference when the issue of a ban on nuclear testing became more political than scientific.[69] But Killian's lasting success was the elegance and cool-headedness with which he steered the administration from unstructured ideas coming from the missile program, the demand for effective reconnaissance, the underfunded efforts of the IGY to a clear separation of military and civilian space programs, and the effective organization for the non-military space program through NASA.

However, by July 1959, the balance between Killian's input and the value he brought to the White House was shifting. As he said himself: "In early spring of 1959, I also became convinced that I had pretty well done the job that I, as a science administrator, had agreed to undertake. In the coming months ahead, a fully-fledged scientist could best help the President in coping with the kinds of problems that were to be on his agenda." Given that this statement comes from Killian's memoir, it could be a matter of retrofitting views to fit the circumstance. But there is an air of logic and self-understanding that makes it easier to agree with Killian. Later in the year though, in his diary, Kistiakowsky, Killian's successor wrote:

> During cocktails, Mrs. Lewis Strauss [wife of the Chairman of the Atomic Energy Commission] made strenuous inquiries as to the real reasons which induced Dr. Killian to resign. It was clear from her re-

> marks that my statement, that it was his wife's health and pressure from MIT was not adequate for her.[70]

Actually, Killian's official letter of resignation said:

> Dear Mr. President,
> For compelling personal reasons, about which you have been warmly understanding, and in order to return to the Massachusetts Institute of Technology, from which I have been on leave, I must regretfully resign my post as your Special Assistant for Science and Technology.[71]

Eisenhower would not have hastened his return to MIT, nor did he go to any great length to persuade him to stay, despite the fact that he enjoyed Killian's company, his insight and his rare ability to translate the complex world of deep science and technology into clear, actionable language that Eisenhower could engage with. Killian received the usual valedictory letter from Eisenhower marking the departure of a subordinate, and reflecting his success with PSAC. The president released the letter to the press, writing: "Through your experience and clear judgment, brought to bear on many complex problems, you have been of inestimable help to me. . . . The work of the Science Advisory Committee, in which I know you have played an effective part, has already produced results that should have lasting value to the nation."[72] Just after Killian left Washington, Eisenhower wrote a rather more personal, off the record (in so far as Eisenhower's writing was ever off the record) "Dear Jim" note. Eisenhower was never truly informal in his written correspondence, but there's a sense of both friendship and gratitude in his writing. He wrote:

> No one did more than you, in those early days, to bring reason, fact and logic into our plans for space research and adventure. I shall never cease to be grateful for the patience with which you initiated me into the rudiments of this new science. . . . More than all this, every contact with you has been, for me, interesting, informative and often inspiring.[73]

This candid friendship and respect is echoed in two other letters Killian kept long after he had left Washington. One, signed "With warm regard, sincerely D.E." is an invitation to Killian to attend one of Eisenhower's "occasional stag dinners."[74] This placed Killian very much within Eisenhower's inner circle. The stag dinners were the president's opportunity to talk informally about subjects of national importance with men whose opinion he valued. The second letter reflects that even the somewhat austere and aloof Eisenhower did make time to recognize the achievements of those he trusted to drive forward his agenda. On April 20, 1959, in another "Dear Jim" letter, Eisenhower wrote:

> From the compliments that have reached me on this year's conference for the Advertising Council and specifically on the panel on science and education that you put together—I judge that this was the best

> received and therefore most useful to the Administration of any that we have held. I am especially grateful to you for your participation as I have learned that because of a cold and some night travel, it was at some considerable personal sacrifice that you did it.
>
> With warm regard, as ever, DE.[75]

It is quite understandable why Eisenhower was so fulsome in his praise for his first science adviser. Killian had fulfilled all the expectations of a trusted lieutenant. He had taken clear instruction from Eisenhower on his appointment and had delivered on those instructions without fuss and without adding to the burden of the presidency. In short, he had been the perfect presidential "Helping Hand."

There is a significant and telling contrast between Eisenhower's close association and the rather more stilted and formal letter Vice President Nixon wrote to Killian as the latter returned to MIT. While this too is a "Dear Jim" letter and signed "Dick Nixon," there is considerably less sense of intimacy, perhaps reflecting Nixon's position on the outer edge of presidential decision making. Nixon wrote:

> You answered an emergency summons to serve your country in an area which demanded the attention of the greatest minds of the scientific community. To an area dominated by public hysteria you brought the calm serenity of objective insight. . . . The debt which this nation owes you will be determined by the future course of history, but you can be assured of the fact that you have rendered invaluable and outstanding service to your country.[76]

One last service Killian provided for Eisenhower was to provide four recommendations as to who should succeed him. Kistiakowsky was at the top of the list which also included Brooks, Bacher, and DuBridge, in descending order. Robert Bacher and Lee DuBridge had worked with Killian since the Office of Defense Mobilization Science Advisory Committee days, while Harvey Brooks, who eventually joined PSAC in 1960, was the dean of engineering and applied physics at Harvard. Bacher, a former AEC commissioner was a professor at Caltech, while DuBridge was Caltech's president, a trustee of the RAND Corporation and former member of the National Science Foundation.[77] Kistiakowsky, a leading chemist in the Manhattan Project and former chair of chemistry at Harvard was Killian's first choice. He had already been an effective member of PSAC since its formation and was well-known to the president. Eisenhower agreed to Killian's suggestion without question.

Kistiakowsky's arrival almost coincided with the completion of NASA's structure. Glennan had control of the Mercury Program, the Saturn booster project, the JPL and was soon to finally wrest the ABMA from the Army's clutches. The focus now for NASA was on Eisenhower's legacy: what sort of space program was his administration going to bequeath to the nation? That largely rested on budget negotiations for

FY1962, and, of course, on the 1960 presidential election and who would succeed Eisenhower in the White House.

THE FINAL YEAR—LEAVING A LEGACY IN SPACE

Eisenhower's last year in office should have been valedictory. Had it gone to plan, he would have left the White House following a successful summit with the Soviets, a nuclear test ban agreement and a balanced budget. In the best scenario, the Republicans would have regained Congress and Richard Nixon would be succeeding him as president. Of course, it was not to be. By the time the administration was preparing its final budget and plans to hand on the winner of the 1960 presidential election, relations with the Soviets had soured over the U-2 incident and failed Paris summit. Hopes of a nuclear test ban treaty diminished by the day as the scientists' initial hopes for a breakthrough were extinguished, and the treaty talks became mired in ideological posturing, much of it within the United States' own armed forces, AEC, and congressional political factions.[78] The raised temperature of the Cold War, following Khrushchev's storming from the Paris conference room, coupled with deteriorating relations with Cuba and the constant sniping of the Democrats on the supposed missile gap, led to a rather more fraught year than Eisenhower would have hoped for, and certainly not the kind of final victory lap a president would want to build a legacy on. During his final months in office, the focus for his legacy became two-fold: leaving the nation at peace, and ensuring the new administration inherited a balanced budget. For NASA's civilian space program, that meant an uneasy balance between ambitious space plans and limited funding with which to realize them. For a Republican candidate on the presidential trail, it meant the even less comfortable balancing act of supporting the current administration while facing the challenges of a seemingly young, vigorous, and radical opponent.

While Eisenhower supported the Mercury man-in-space program, he was far less enamored with the idea of sending a man to the moon. While he had supported extra funding for NASA to acquire the Saturn booster project and the ABMA team that was developing it, the first the president knew of NASA's active plans to send a man to the moon was when the program was presented to him in January 1960. Kistiakowsky noted that at a January 8, 1960 meeting, Eisenhower agreed to "resolve conflicts between DoD and NASA regarding booster development and similar issues." It would seem that he was not expecting any internal rivalry issues, but was keenly aware of reports of continuing Soviet progress in missile development.[79] The president had always acknowledged since the Sputnik launch that the one US deficiency in space was a powerful rocket booster. It had been unnecessary to develop such a booster earlier in the 1950s when the focus had been on developing workable ICBMs. But with the options for larger and more powerful satellites and both unmanned and manned scientific space stations, as laid out in Purcell's 1958 *Introduction to Outer Space* report, the president was a strong advocate of prioritizing the large booster program. While Saturn appeared to offer little to the United States' ongoing military development, according to Kistiakowsky: "at the very end [of the meeting, Eisenhower] was rather favorably considering the idea of adding another $100 million to the Saturn project if that could speed it up."[80]

At the final Space Council meeting under Eisenhower's presidency on January 12, which was attended by Nixon and his future running mate, Henry Cabot Lodge, as well as most regular national security committee attendees, Glennan presented the long-range plan for NASA. For the first time, this talked to an audience beyond NASA insiders about a follow-on man-in-space program after Mercury, and offered a mission to the moon as part of the raison d'être for the Saturn super booster. Kistiakowsky noted that this: "was followed by an emphatic statement by the President that maximum effort should go into super booster rockets. [Maurice] Stans [Director of the Bureau of the Budget] objected to the expenditure of so much money and was firmly dressed down by the President."[81] At this stage, Eisenhower was not endorsing a moon landing program. He saw far more benefit in unmanned over manned space exploration and did not equate the Saturn booster with a manned landing on the moon. Indeed, his August statement on US successes in space and his final discussions over the budget both pointed to a strong resistance to endorsing manned space efforts beyond the Mercury proof of concept. In one area of that August statement, there was no change at all. Eisenhower's long game of satellite development based primarily on national security needs was not driven by the same propaganda agenda that drove Khrushchev, and nor was it a response to the provocative nature of the Soviets' series of "spectacular first" efforts. Speaking about the satellite successes, he said:

> All these are the results of a well-planned and determined attack on this new field—an attack that promises very real and useful results for all mankind. Each of these satellites is destined to play an important part in broadening man's understanding of the cosmos in which he lives. While no one of them has been undertaken solely in an effort to achieve a "spectacular first" in the eyes of the world, each has resulted in just such a "spectacular first" in support of the desires of mankind for greater knowledge and understanding.
>
> The United States leads the world in the activities in the space field that promise real benefits to mankind.[82]

The "real benefits" were scientific—discoveries such as the Van Allen Belts; and technological, making strides in meteorology and communications. And, of course, Discoverer/CORONA heralded the step away from the reliance on U-2 led reconnaissance to the new age of the space satellite. Man-in-space did not fit this pattern. Eisenhower saw the Saturn program as giving NASA a good reason to draw in the talents of the ABMA team, and the resulting booster would pragmatically fill the heavy booster gap that the Soviets had opened up in 1957. There was no romance in funding Saturn; this was not a prestige project. It may simply be that he saw the Saturn booster as the launch vehicle for future space stations, a concept presented by NASA in 1959 following Purcell's initial study, or that the pragmatist president saw a potential cross-over into military use further down the line in Saturn's development. However, he did endorse NASA's long range plan which was shared with Congress on January 26, 1960. At that stage NASA's moon mission plans were quite vague, which probably made it easier for Eisenhower to endorse the overall ten-year strategy. A great believer in plans, he never expected them to transpire in quite the way they were laid out on paper. NASA introduced the moon mission concept by saying: "In the long run, such activities [a broadly and soundly conceived program of research, development and operations in space] should make feasible the manned exploration of the moon and nearby planets and this exploration may be taken as a long-term goal of NASA activities."[83] That was a long way from talking overtly about the cost and procedure of landing a man on the moon.

While endorsing the plan at a high level, Eisenhower could constrain it while still in the White House, by limiting the funds NASA had to work with. And, as he entered his last round of budget negotiations, he showed no signs of loosening the reins on spending on what he perceived as spectacular rather than substantive space projects. But that final round of budget planning was still some months off. What would dominate the domestic agenda for 1960 was not any "space race" but the race for the White House and the election to succeed Eisenhower as president.

NIXON AND SPACE

Richard Nixon had never been a key player in developing space policy. While his vice president's role had given him the Senate chair and, theoretically, ensured he was Eisenhower's link with Congress, the president had chosen to bypass Nixon and deal directly with Lyndon Johnson in the post-Sputnik period through to the passing of the Space Act. Nixon, through most of Eisenhower's presidency was the outsider looking in. The U-2 incident on the eve of the Paris summit had initially damaged the Republicans, but counter-intuitively, Eisenhower had emerged stronger from it. After the weak cover stories that asserted that Francis Gary Powers's plane was nothing more than a NASA weather flight had been shredded by a gleeful Khrushchev, the president had wrested the initiative by acknowledging the true nature of the flight, and refusing to apologize to the Soviets for the United States' show of power in having an active four-year reconnaissance project that had already delivered powerful intelligence via the CIA's U-2 flights. Again, Eisenhower showed he could change tack if the circumstances demanded. Previously, during the winter of 1958–1959, Eisenhower had chosen to keep his counsel as Khrushchev provoked him over Berlin. The accusations from both the media and, particularly, from Democratic politicians were that Eisenhower was weak in the face of Soviet aggression.[84] Eisenhower had surmised correctly that Khrushchev's threat was bluff. With the U-2 incident, the tables were somewhat turned. Eisenhower demonstrated that the United States was not inactive in the face of the Soviet threat but was, and had been for some time, actively countering the Soviets by means of cutting-edge reconnaissance. This had always been risky, but it was a risk Eisenhower accepted. Deputy CIA Director Robert Amory had told Eisenhower from the outset that the U-2 had a limited life. "The Russian radar would improve, their fighters and intercepts and other things like their surface to air missiles would improve. And a precisely accurate prediction was made of about a four-year life."[85] By standing up to Khrushchev, even in the uncomfortable light of the drawn-out humiliation the Soviet leader tried to impose, slightly perversely, Eisenhower regained the respect and support both of the US public and the wide free world. Indeed, in a Gallup poll taken in June 1960 after the furor around the incident subsided, 58 percent of Americans questions stated that the president had handled the U-2 incident well.[86]

This really did not help Nixon as he prepared his run for the presidency. By the summer, as the president began a final world tour, Eisenhower's stock was on the rise again. Immediately following the failed Paris summit, Eisenhower's approval rating had fallen to 49 percent. By September, it was back to 65 percent.[87] Running for the White House, Nixon faced a double challenge in attempting to appear distinct to the electorate. He was torn between his duty to support the exiting president's insistence on leaving office with a balanced budget, and his own political desire to confront Kennedy with a more aggressive and perhaps more populist agenda. One area where he could take on Kennedy from a position of strength was on achievements in space. Yet he was limited in how far he could push his own agenda by having to appear continually supportive of both Eisenhower's handling of the U-2 incident and the president's insistence in avoiding a space race.

"Highlights in Space Exploration" was a briefing paper prepared for Nixon by John Hamlin who had left the White House staff to work for Nixon as a researcher and speechwriter.[88] Hamlin's two-page document recorded a succession of US satellite successes. Explorer 1-VII and Pioneer 1-V detailed a string of scientific successes. Vanguard revealed the earth was "pear shaped."[89] Tiros 1 was the first "picture-taking weather satellite" while Transit 1B and 11A were the first time that two instrumental satellites were placed in orbit at the same time. Echo 1 was the largest artificial object in space, while the X-15 "set new records for speed (2196 miles per hour) and new record for altitude (136,500 feet)."[90] Perhaps the most interesting reference was to Discoverer XIII and XIV. These were described thus: "Brings nearer the goal of manned space flights by successfully recovering satellite capsules ejected from orbiting satellites."[91] What was most certainly not explained was that these were tests of the CORONA spy satellites: "black" projects hidden in plain sight under the "white" banner of NASA scientific flights. What should have given Nixon the greatest satisfaction, and the most significant evidence that Eisenhower's space policy was working, was the "Balance Sheet" that closed the document. It read:

Table 5.1.

	United States	Soviet Union
Number of space vehicles successfully launched	26	8
Number still in space	16	3
Number still sending back data	9	1
Number of space capsules recovered from orbit	2	1

"Highlights in Space Exploration," Sections 60-92, Box 4, Space & Aeronautics, Rockets & Satellites, Pre-Presidential Papers, 1960 Campaign, Hamlin File, Richard M Nixon Presidential Library, p. 2.

However, Nixon had little opportunity to press the case that the US space policy was a success as the Democrats changed the focus of debate, mounting the attack not on Eisenhower's civilian space policy which was developed and executed in plain sight (with the exception of the hidden role of the Discoverer/CORONA program), but on the so-called missile gap. There was indeed a missile gap: probably 4–1 in favor of the United States (see table 5.1).[92] But Eisenhower was unbending on his stance of never making national security intelligence public. In being rigid on this policy, he undermined Nixon's chances of winning the election. Eisenhower did make one statement on space policy during the campaign, although the comment on the success of US efforts in space released at his news conference on August 17, 1960, sought to play down any inference that the United States was interested in a race in space or in delivering "spectaculars" along the lines of the Soviet Sputniks. For Eisenhower, space exploration could be exciting, but it was a means to an end. His policy was successful, but the meteorological, communication, and navigation satellites it was delivering were functional rather than spectacular, and the Mercury manned program was a proof-of-concept, rather than any kind of pioneering adventure.[93] It did not offer Nixon an opportunity to outflank Kennedy.

Of course, it was much more than small differences from Eisenhower in space policy that really stymied Nixon. The president had chosen to remain largely aloof from the campaign, especially in its early stages. And his sense that he was always the best person to deal with any significant policy decisions reflected badly, inadvertently, or otherwise, on the public perception of how he viewed Nixon. This is not the place to reflect on Kennedy's victory, but the fact that a Democrat would take over the White House in January 1961 did affect Eisenhower's final actions in the winter of 1960–1961.

RECALIBRATION

Given that he had actively supported additional funding for the Saturn booster in 1959 in compiling the FY1961 budget for NASA, it may be surprising that Eisenhower had apparently lost confidence with NASA's ability to implement his strategy a year later. Yet it should really come as no surprise. Eisenhower could not support NASA's increasing concentration on follow-on man-in-space developments after the conclusion of the Mercury program. It demonstrated an executive agency deviating from the executive strategy without any direct order to do so from the president. Eisenhower's ambivalence bordering on hostility to man-in-space became clearer late in 1960 when NASA submitted its budget request for FY1962 to the Bureau of the Budget. The application detailed, for the first time in public, NASA's request for funding for a lunar landing program. This was most certainly not what Eisenhower had envisaged. Meeting with the president in October, Glennan noted:

> At 3 o'clock, I met with the President. I found him [Eisenhower] tired and preoccupied. I simply brought him up-to-date on our activities . . . and then indicated that we were going to require additional money in the way of a supplemental. He had no comment to make on this matter. I told him something of the costs that appear to be involved in Project Apollo, the follow on to Project Mercury. He expressed himself once more as having little interest in the manned aspects of space research. He was cordial enough but it was obvious that he was not at his best today.[94]

Glennan clearly failed to get the measure of Eisenhower at the October meeting. While the president had been confident to let Glennan and NASA implement his strategy through the previous eighteen months and therefore refrained from extensive personal involvement, there were now clear signs that the president's view of the strategy and NASA's interpretation of it were diverging. NASA was already planning to send a man to the moon, while Eisenhower did not view man-in-space as a priority for the nation. The consequence was that Eisenhower returned his active attention to NASA. He asked Kistiakowsky to study: "the goals, the missions and the costs" of NASA's planned program. Kistiakowsky set up a six-man panel led by Donald Hornig of Brown University to conduct the study. They reported back on December 16, 1960. The report opened quite portentously—not the usual PSAC style—as it stated: "We have been plunged into a race for the conquest of space . . . the most compelling reason for our effort has been the international political situation which demands that we demonstrate our technological capabilities if we are to maintain our position of leadership."[95] Such rhetoric never won favor with Eisenhower, and perhaps reflects a too-close relationship between Hornig's team and their NASA interviewees. Indeed, the report went on to praise NASA's co-operation: "Officers of the NASA presented a very impressive description of their detailed plans for development, utilization and costs of the Saturn vehicle." But over the remainder of the report, the team reverted to a cool appraisal of NASA's aspirations.

This was the use of a technical PSAC panel in the way that Eisenhower preferred: essentially telling him what he suspected, but needed to be validated by scientific professionals. Project Mercury was described as a "marginal effort" which owed its continuation only to "the political desire either to be the first nation to send a man into orbit, or at least be a close second."[96] The Saturn C2 program, as currently envisaged, would enter its test phase about 1965, and was "expected to lift about 40,000 lbs. into low earth orbit and planned to utilize this capacity to send up an 'orbiting laboratory.' . . . It is our opinion that an orbiting laboratory of this size could produce considerably more scientific information if it were wholly instrumental rather than manned."[97] The report noted that the planned Saturn developments and Apollo spacecraft were not "capable of landing on the moon with sufficient auxiliary equipment to return the crew safely to the earth"[98] and indicated that a new program much larger than Saturn would be needed. The Nova program was discussed alongside a series of unmanned programs.[99] The response of the scientists to the comparison was unemotional, but touched on the reasons for manned space exploration that Eisenhower's cold rationality simply did not engage with. "Certainly among the major reasons for attending the manned exploration of space are emotional compulsions and national aspirations. . . . It seems therefore to us that man-in-space cannot be justified on purely scientific grounds."[100] Finishing the report with the likely costs, manned earth orbit was estimated at $350 million. Manned circumnavigation of the moon was put at $8 billion. A manned lunar landing ranged between $26 billion to $38 billion. Hornig's team concluded: "The unmanned program is a necessary prerequisite to a manned program. Even if there were no manned program, the unmanned program might yield as much scientific knowledge and on the basis would be justified in its own right."[101] Four days later, Glennan recorded his attendance at an NSC meeting at which Project Apollo was discussed.

> I presented our budget and 10-year plan as revised while Hugh described in more detail the activities to be undertaken under the 1962 budget, which is now set at $1.16 billion approximately. Of that amount, $50 million will be requested in a supplemental for the current fiscal year. After we had completed our story showing the NASA budget increasing to more than $2.5 billion annually by the end of the decade, Kistiakowsky talked about the manned space flight program beyond Project Mercury. Most of his information had been derived from presentations given by our people to a committee of the President's Science Advisory Committee. The total dollars estimated to be required for landing a man on the moon and returning him to earth are really quite staggering. One can support a figure anywhere from $10 billion to $35 billion and even then, not know whether or not he is in the right ballpark.
>
> The President was prompt in his response: he couldn't care less whether a man ever reached the moon. There was desultory comment by others in the meeting who were concerned over the increasing cost of space research. . . . In some ways, the meeting was discouraging. However, I think that feeling might be considered a natural one under the circumstances.[102]

Eisenhower was extremely dismissive of the NASA plans. NASA Associate Administrator Robert Seamans recalled the meeting saying: "Eisenhower ended this part of the meeting with a rhetorical question: "Can anybody tell me what is the best space program for $1 billion?" Walking from the cabinet room, I realized why Maury Stans was adamant that there would be no additions to NASA's budget in FY 1962."[103]

By January, Eisenhower's view on the manned space program had hardened to the extent that he wanted to speak out against it in public before leaving the presidency. Glennan recorded:

> Dryden and I visited Kistiakowsky to attempt to get a change made in the budget message of the President, which is to be delivered on 18 January. Apparently, following the National Security Council meeting on the 20th, a statement was prepared for inclusion in the message that would, in my opinion, be unwise. The President proposes to say that there is no scientific or defense need for man in space beyond Mercury. It is much better, if I am any judge of the political realities of the situation, to say that we need much more research and development before a definite decision can be made in this matter. Actually, such a statement would be in complete agreement with the facts as they will be presented in the budget message. After much telephoning, we were able to get this statement changed.[104]

The budget speech was delivered without the passage stating that there was no scientific or defense justification for continuing the man-in-space program beyond Mercury, but in his final weeks in office, Eisenhower did instruct Goodpaster to circulate the Hornig *Report of the Ad Hoc Panel on Man in Space* report to Congressional leaders and anyone he thought might have an influence on future NASA policy.

Nor was Eisenhower alone in raising objections to NASA's manned lunar landing plan. Surprisingly, some of the criticism came from within NASA itself. As late as May 21, 1961, Hugh Dryden, NASA's deputy administrator, testified before the Senate Appropriations Committee. When asked what practical use he saw in landing a man on the moon, he responded: "It certainly does not make any sense to me."[105]

THE HAND-OVER PROPOSITION

As part of the transition to the new administration, PSAC member Jerome Wiesner chaired an ad hoc committee on space producing a report for the president-elect on January 10, 1961.[106] Wiesner, soon to be appointed Kennedy's special scientific adviser, did not call on his predecessors Killian or Kistiakowsky to help compile the report, but did include Ed Purcell, author of the 1958 report that laid the foundations of NASA's structure and operations; Trevor Gardner, the Air Force's special assistant for research and development; Din Land, the Polaroid president and mastermind behind the camera technology on both the U-2 and CORONA programs; and Donald Hornig, a member of the National Academy of Science and Wiesner's eventual successor as the presidential science adviser. Wiesner, a registered Democrat, had worked with the Democratic Party throughout the 1960 presidential election campaign, a point that had not gone unnoticed in the White House. As early as July 12, 1960, Kistiakowsky recorded a meeting of PSAC with the president at Newport, and a follow-up conversation with Wiesner who was also present at the meeting. Kistiakowsky noted he told Wiesner he: "should influence his friends, who in turn influence Kennedy, to have the latter pick out a special assistant for science and technology early enough so that as soon as the election is over, the man can start getting acquainted with our work." Wiesner indicated he had already been told he was Kennedy's choice. Kistiakowsky was absolutely not conceding the election this far out, but was being realistic that the next administration could be Democratic. He also recorded a private conversation with Killian urging him to work for Nixon during the campaign "as a counterweight to the influence of Teller."[107] On August 22, White House Counsel David Kendall contacted Kistiakowsky suggesting: "that in view of Wiesner's involvement in the Kennedy campaign, he should be excluded from PSAC activities until after the election."[108] Kistiakowsky responded to Kendall that PSAC was non-partisan and that half its members were Democrats—and was able to convince Kendall that Wiesner should stay an active member "providing Wiesner's political activities are non-public."[109] A month later, Wiesner himself called Kistiakowsky to inform him that he (Wiesner) was "getting more and more alarmed by some of the things Kennedy is saying, obviously under the influence of the advisors around him, like Symington, Murray etc."[110] He asked what the White House reaction would be if he "joined Kennedy's party for a while." Kistiakowsky told him to follow his conscience but noted: "I recall speaking about Wiesner's activities with the President and his agreement, after some hesitation that Jerry remain on the PSAC so long as he does not publicly criticize the Eisenhower administration."[111] Wiesner was finally appointed as special assistant to President Kennedy for science and technology and, simultaneously, as chairman of the President's Science Advisory Committee, following the inauguration. Thus his report to Kennedy on space came before he was officially in role.

Having been critical of Eisenhower's passive and active defense policies through the Gaither Panel, of which he was technical director, one might have expected Wiesner's transition report on space to be highly critical of Eisenhower's 1958–1960 space policy. Yet that is not the case. While the report is critical of NASA's management and makes a recommendation to "Establish a single responsibility within the military establishments for managing the military portion of the space program,"[112] the focus of the report is on improving organization and integration to build on the previous two years' experience. This was not a radical report. It was not the rallying cry for a crash man-to-the-moon program that one might expect given Kennedy's pledge to send a man to the moon addressed to Congress just five months later. Indeed, it is far from it. The most pressing recommendation, though noted as the fourth out of five, hand-amended by Wiesner, was as follows:

> Review the national space program and redefine the objectives in view of the experience gained during the past two years. Particular attention should be given to the booster program, manned space flight, the military uses of space and the application of space technology to the civilian activities of the country.[113]

This was not significantly different from Nixon's platform and much more a continuation of the Eisenhower strategy than what actually followed under Kennedy. It differed from Eisenhower in that there was a drawing together of military and civilian activities, but shared the conclusions of Eisenhower's own commissioned report in questioning the value of manned space flight. In fact, Wiesner's report was rather dismissive of manned flight. While it is given its own section, "man in space" is actually the seventh topic discussed, coming after a general review and then sections on ballistic missiles, organization and management, the booster program, the military space program, and science in space. While Wiesner recognized that "We are rapidly approaching the time when technology will make it possible for man to go out into space. It is sure that as soon as this possibility exists, man will be compelled to make use of it."[114] Yet the report regarded military and scientific missions as only "dimly perceived,"[115] while an underlined comment noted: "It is very unlikely that we shall be the first in placing a man in orbit around the earth."[116] There was almost a grudging note when the report, which had devoted twice as much coverage on space exploration to unmanned rather than manned missions states:

> While the successful orbiting of a man about the earth . . . will provide a necessary stepping stone toward the establishment of a space station and for the eventual manned exploration of the moon and the planets . . . a crash program aimed at placing a man into orbit at the earliest possible time cannot be justified solely on scientific and technological grounds. Indeed, it may hinder the development of our scientific and technical program and even the future manned space program by diverting manpower, vehicles and funds.[117]

Thus, Wiesner's report prepared to enable the new President to hit the ground running was not a new Democratic strategy setting sail on the oceans of space toward a new frontier as Kennedy would talk about at Rice University in 1962. This was a far more pragmatic request to improve organization and prioritization to make incremental improvements to a strategy borne out of the twin developments of missile R&D and the scientific investigations initiated under the IGY. As such, the planned Wiesner program as outlined in the report was far closer to the Eisenhower legacy than one might expect from a Democrat.

Given the significant shift in support for the man-in-space program that was to take place under Kennedy, it may be surprising that Wiesner was so lacking in support for manned space exploration. But the explanation is simple. Although he was a Democrat, Wiesner was first and foremost a part of PSAC's inner core. PSAC was non-partisan and had been created to give the president unbiased scientific expertise: advice based on scientific knowledge, not political considerations. By 1960, it had reached its political apogee. But was it quite as unbiased as it seemed? Eisenhower had identified Killian, initially through his work with the TCP, as someone who shared his general values and political outlook. Killian was driven by these values in selecting PSAC scientists, thus rejecting those such as Teller and Von Braun who sought to further their own agendas rather than that of the president. Therefore, the PSAC panels became collegiate entities grouped around Eisenhower's basic philosophies. It is pushing the analogy too far to say there was any kind of "groupthink" in their decision making.[118] However, they knew the principles of the president and had a tendency, on space especially, to deliver verdicts in line with those principles. The lack of success of the Gaither Panel in 1958 in changing the president's perspective on national defense gave a clear message to PSAC on what the president's expectations were from the team he called "my scientists." Killian captured the bond between the president and PSAC well in his memoir recording his final meeting with Eisenhower shortly before the former president died.

> The general seemed to welcome the opportunity to talk, and the visit lasted for nearly an hour. . . . At one point, he asked about "my scientists" and specifically mentioned several by name. Then he made a comment that I shall always cherish: "You know Jim, this bunch of scientists was one of the few groups that I encountered in Washington who seemed to be there to help the country and not themselves." His statement was true.[119]

Don Hornig, another Democrat, who became Johnson's Scientific Adviser, was questioned about PSAC's relationship with Eisenhower in an oral history interview conducted in 1968.

> The Science Advisory Committee covers everything from health to military intelligence. On any specific problem the approach has been to set up a panel or task force. The notion is that a group works on a major problem, works intensively for a year and works as part of the Presidential family. They are outsiders, but in fact they are given very high security clearances, made reasonably privy, to the internal pullings and haulings of the government, at least in the areas with which they are concerned. They have brought in objectivity from the outside, but they have also been insiders in the sense that they have been part of the White House family.[120]

Being part of the "White House family," as Hornig explained, was important for building the symbiotic relationship developed between Eisenhower and PSAC. Equally important was the mode of working through panels and task forces.

> Starting with the pattern that Killian set, the kind of task forces we've had in our office have been enormously useful . . . from the beginning we didn't identify publicly the existence of taskforces; we refused to divulge publicly the membership of task forces, panels of the President's Science Advisory Committee. The reason was just to give people a chance to work on problems without being exposed to political pressures personally. For instance, the members of the President's Science Advisory Committee, I figure, on the average spend thirty-five or forty days a year working for him. Considering that they are all top-notch people, this derives from the fact that they think they are effective and they think it is worthwhile.[121]

What was most effective and most worthwhile for Eisenhower was that these panels and task forces were not subjected to political interference, and nor were they running the gauntlet of lobbyists from the "Military

Industrial Complex." They could assess the evidence coolly and come to viewpoints less affected by political influence. Therefore, to a greater degree, the advice Eisenhower received from PSAC was "pure." Additionally, this was not a one-way street. As one saw in the months after Sputnik, Eisenhower did not force his opinion on how the civilianized aspect of the US space program should be managed. While his initial intention was for it to come under military control, he was persuaded by Killian's PSAC experts that the best route was a civilian agency. Thus it is wrong to see the Eisenhower/PSAC relationship as one of command/deliver. While it was framed by the strategic principles of the president, notably economic conservatism, national security, and avoiding conflict, it was a relationship of mutual respect although it exhibited slight deference to the president. As with that inadvertent slip at the August press conference that gave the Democrats their "If you give me a week" ammunition, it was always clear that Eisenhower would make the final decisions on issues raised and researched with PSAC. Their role was to advise. His was to lead. Speaking at MIT in October 1973, James Killian summed up the success of the relationship:

> The importance of PSAC goes far beyond the specific outcomes of its studies and recommendations because of the relationships of confidence and free discussion that PSAC enjoyed with the President and the President's associates. These meetings, in which there was free-for-all discussion, were memorable events for PSAC itself. They made it possible for a group of scientists to come to understand the President's problems, views and goals and to learn how to make themselves useful in the light of this understanding. So it was that the committee found many ways to express its belief in the values of a free society not only for the advancement of science, but for the good of mankind.[122]

The stance taken by PSAC squared firmly with Eisenhower's own sense of duty to the country which elevated his attitude toward executive decision making above partisan politics. Kennedy was to use the space program to make political capital at the expense of a long-term strategy. In so doing, he largely abandoned the fledgling but logical foundation that Eisenhower had set for the scientific exploration of space.

THE EISENHOWER SPACE LEGACY

Ken Collier coined the term "Autopilot Presidency" to describe Eisenhower's relations with Congress, where he defined Eisenhower as a leader who stepped in only when it was necessary to deal with turbulence.[123] However, to apply that same analogy to the way the president led on space policy once NASA was operational is not quite accurate. The analogy is close, but needs building further. To extend the flight analogy, Eisenhower picked the destination and set the course. He flew the plane to its cruising height, and then handed over the operational flying to his first officer. When any difficult course changes were necessary, and when it came to landing the plane, Eisenhower took control once again. He dealt with more than just turbulence.

On leaving the White House in January 1961, Eisenhower could be proud of the edifices he had created and relationships he had built in support of the implementation of the United States' first space policy. In twenty-seven months, he had formally created and mandated a science advisory committee that put objective science into the heart of executive decision making. Although PSAC's remit was far wider than space, and was heavily focused on nuclear test ban talks, basic science in education, missile development, chemicals in food, and high energy particle accelerators among other projects; arguably its finest achievement was to shape the Space Act in 1958.[124] This Act enabled the creation and operation of NASA, a civilian agency reporting directly to the president. The very creation of an arms-length agency reporting to the president headed off the rivalry between the armed services in the pursuit of primacy in the US space program. Of course, this rivalry bubbled in the early years of NASA's existence, never more so than in 1958 where Medaris and, to an extent, Wernher Von Braun, fought a rearguard battle to preserve the ABMA's independence within the Army, beyond the grasp of NASA control. Equally, the Air Force's backing for military space spectacular projects such as Boeing X-20 Dynasoar which ran from 1957 to 1963 caused friction with its civilian peer and drained significant funds (over $660 million in the case of Dynasoar) at the precise time that NASA was designing follow-on projects for the proof-of-concept man-in-space Mercury program.

Part of Eisenhower's legacy is that he placed the Mercury program in the hands of NASA. But, by 1960 with the relative success of US satellite launches in comparison to the Soviets; the rapid advance in the development of the US IRBM and ICBM packages; and the first payload returns of the CORONA spy satellite; his enthusiasm for man-in-space, if it had ever existed, had waned. It is hardly surprising since the likelihood still was that the Soviets would launch the first man in space, and Eisenhower had repeatedly made it clear, not least in his space speech of August 1960, that he did not believe in the merit of space spectaculars. This was not what he had created NASA for. This was the scientific arm of space discovery, running in parallel with the military space program and not the platform for a Cold War battle. The NASA he left behind was drawing together the best engineering minds in the country through its network of internal personnel and rapidly-growing engineering technology partners. But it was also growing a sense of self-importance that placed its ambitions and goals beyond the limited role the president saw for it.

In allying with Killian, Kistiakowsky, Purcell, York, and with first NASA Administrator Keith Glennan, Eisenhower had stuck by his principles of believing in small government and in acting in the best interest of the nation, as far as possible, outside the scope of partisan pressure. But, as his farewell address showed, he was concerned that his legacy was in danger of becoming both economically unhealthy, and militarily dangerous.

With a planned program of communications, navigation, science discovery, and meteorology satellite launches, planetary probes, and proof-of-concept man-in-space work in the pipeline, Eisenhower's space program was modest, but economically sound and set for growth toward space stations and longer-range exploration through the Saturn booster, which Eisenhower also supported. It was a conservative program, restrained, logical, and coolly scientific. This was not a legacy of hyperbole and, as such, was a perfect reflection of the president.

Eisenhower's legacy was not lost through Kennedy's moon pledge speech. It took the new president's assassination and a wave of emotional endeavor, championed by Johnson, to see Kennedy's moon pledge brought to reality. By that time, Eisenhower himself was dead. He died just three months before Neil Armstrong stepped onto the moon's surface, never having changed his position on the folly of a manned mission to the moon. There is a certain irony that the man on the end of a telephone making a long-distance call to the moon was another new president: Nixon.

It is impossible to judge whether Eisenhower's slow but steady, moderate approach to civilianized space exploration would have achieved as much or more than the approach taken by his successors. However, at least one man who benefitted from Kennedy's push for the Apollo program has questioned whether it was the right thing to do. In 2009, I sat drinking a beer with Charlie Duke, Capcom to Apollo 11 and the Lunar Module pilot on Apollo 16. I asked the tenth man to walk on the moon if NASA had got things right. "Hell no," he responded. "We did things all in the wrong order. We should have built the shuttle first; then assembled a space station. That would have given us a great jumping off point to the moon and far beyond. Kennedy killed those plans and put us back decades."[125]

NOTES

1. In the specific works written on Sputnik, Divine's *Sputnik Challenge* concludes in 1958. Dickson's *Sputnik: The Shock of the Century* does cover the remainder of Eisenhower's presidency, but he relies solely on second-hand source material and uses Eisenhower only as a poor comparison to the activism Kennedy would bring to the "space race." Mieczkowski, using more detailed interview transcripts and primary source material actually takes a similar line in *Eisenhower's Sputnik Moment,* and the focus of the last third of his research is not on Eisenhower, but on the presidential election race of 1960 and Kennedy's space strategy.

2. Public statement of DDE on the death of John Foster Dulles, released from Gettysburg, PA, on May 24, 1959. Sourced from "Dwight D Eisenhower, Public Papers of the Presidents of the United States," Office of the Federal Registrar, National Archives and Records Services, (Washington, DC, 1959), p. 117.
3. MA Guhin, *John Foster Dulles, A Statesman and his Times* (New York, 1972) pp. 221–42.
4. A Goodpaster, "Memorandum of Conference with the President," July 17, 1958. Present: Killian, Goodpaster and DDE, Box 35, Staff Memos 1958, DDE Diary Series, DDE Library. The bill became law just twelve days later.
5. Ibid.
6. L Thomas and E Jablonski, *Doolittle, A Biography* (New York, 1976), p. 127.
7. A Roland, *SP-4103, Model Research* Volume 1 (NASA History Office, Washington, DC, 1985) p. 285.
8. JR Hansen, *Spaceflight Revolution: NASA Langley Research Center from Sputnik to Apollo*, NASA History Series, (Washington, DC, 1995), p. 12.
9. Killian, *Sputnik*, p. 139.
10. A Goodpaster, "Memorandum of Conference with the President," July 17, 1958. Present: Killian, Goodpaster and DDE, Box 35, Staff Memos 1958, DDE Diary Series, DDE Library.
11. Doolittle's biographers, L. Thomas and E. Jablonski, note in *Doolittle, A Biography* (New York, 1976) that he was a Director of the Shell Oil Company from 1946–1967, and additionally became Chairman of the Board of Space Technology Laboratories Inc. in 1959. STL was a division of TRW, which developed the Atlas and Titan ICBMs.
12. Hansen, *Langley*, p. 12.
13. Glennan, *Edited Diary*, p. xxi.
14. Glennan, *Edited Diary*, p. 2.
15. Ibid.
16. Ibid.
17. NACA, Memo for files, Faget, Attendance at ARDC Briefing on "Man in Space" Program, March 5, 1958. See also memo to NACA, Soulé, Second Discussion of ARDC Briefing on "Man in Space" Program, March 27, 1958. NASA Organization and Early History Collection, DDE Library.
18. James M. Grimwood, *Project Mercury: A Chronology* (NASA SP-4001, Washington, DC 1963), p. 17.
19. US Air Force "Man in Space Program," March 1958, Folder 18674, NASA Historical Collection, Washington, DC.
20. HT Wells, ST Whitely and CE Karageanes, *SP-4402 Origins of NASA Names*, (NASA, Washington, DC, 1975), p. 105.
21. Glennan, *Edited Diary*, p. 9.
22. Ibid, pp. 10–15.
23. PSAC, "Introduction to Outer Space," pp. 10–15.
24. BR Duchin, "The Most Spectacular Legislative Battle of the Year: President Eisenhower and the 1958 Reorganization of the Department of Defense," *Presidential Studies Quarterly*, Vol. 24, No 2, Eisenhower and Governance, (1994) pp. 243–62 and 1958 State of the Union, Section 4.1, Online Speeches Collection, DDE Library, http://www.eisenhower.archives.gov/all_about_ike/speeches/1958_state_of_the_union.pdf. Accessed September 9, 2014.
25. President's News Conference, April 2, 1958, News Conferences File, DDE as President 1953–1961, DDE Library.
26. DDE, "Address to the American Society of Newspaper Editors and the International Press Institute," April 17, 1958, Washington, DC, Online by Gerhard Peters and John T. Woolley, The American Presidency Project, http://www.presidency.ucsb.edu/ws/?pid=11357. Accessed September 9, 2014.
27. DDE Library, General File, Box 111 contains several examples of sample letters.
28. Glennan, *Edited Diary*, p. 8.

29. Senate voting on HR. 12591. Department of Defense Reorganization Act of 1958. Passage of Senate version http://www.govtrack.us/congress/votes/85-1958/s240. Accessed September 9, 2014.

30. In parallel to the battle fought over defense reorganization, Eisenhower fought an eight-year battle to rationalize US intelligence services. This is eloquently covered in PHJ Davies's first volume of *Intelligence and Government in Britain and the United States,* (Santa Barbara, Denver and Oxford, 2012). Chapter 8, "Ashen Legacies, 1950–1960" opens with Eisenhower's comments to the NSC in the final weeks of his presidency:

> The President also noted that the US fought the first year of the war in Europe entirely on the basis of British Intelligence. Subsequently, each military service developed its own intelligence organization. He thought this situation made little sense in managerial terms. He had suffered an eight-year defeat on this question but would leave a legacy of ashes for his successor.

"Memorandum of Discussion of the 473rd Meeting of the NSC," history.state.gov/historicaldocuments/frus1961-63v25/d80. Accessed September 9, 2014.

31. DDE, "Remarks at the United Republican Dinner and Speech to American Society of Newspaper Editors and the International Press Institute," Box 5, Arthur Larson Papers, DDE Library.

32. MJ Neufeld, *Von Braun,* p. 343.

33. Glennan, *Edited Diary,* p. 10.

34. Kistiakowsky, *Scientist in the White House,* p. lvi.

35. Glennan, *Edited Diary,* p. 10.

36. Ibid.

37. *Baltimore Sun,* 14 October 1958, Section 1, p. 1.

38. J Van Nimmen, L Bruno, and R Rosholt, *NASA Historical Data Book, 1958–68,* NASA Finances, pp. 115–21 (Washington, DC, 1976).

39. Many other newspapers across the country ran the story on October 15, 1958. They included the *Raleigh Register* in North Carolina, which led page 1 with the Associated Press syndicated (but un-by-lined) story: "Army Chiefs Oppose Missile Talent Transfer," while in California, the *Bakersfield Californian* featured Merriman Smith's widely syndicated UPI agency story titled: "Scientists' Transfer Plan Angers Army."

40. Mieczkowski, *Sputnik Moment,* p. 175.

41. "Statistics on Radio and Television," 1950–1960, UNESCO Special Report, (Paris, 1963), p. 23.

42. *The Big Picture: Army Satellites,* April 1958. http://www.youtube.com/watch?v=OjB3larcADw curated by the Air Force Space and Missile Museum, Cape Canaveral, Florida. Accessed September 6, 2013.

43. Neufeld, *Von Braun,* p. 341.

44. NASA, Propulsion Staff, "A National Space Vehicle Program: A Report to the President," 27 Jan. 1959, Johnson Space Centre files, NASA Historical Collection, (Washington, DC).

45. Glennan, *Edited Diary,* p. 22.

46. Neufeld, *Von Braun,* p. 342.

47. Glennan, *Edited Diary,* p. 22.

48. Ibid.

49. BJ Guthrie, under the Direction of RR Roberts, *Statistics of the Congressional Election of November 4 1958,* US Government Printing Office, (Washington, DC, 1959), p. 46.

50. GB Kistiakowsky: Transcript of diary, 1959–60, Box 1, entry for September 29, 1959. DDE Library. [Note: DDE references refer to the unedited manuscript of Kistiakowsky's diary].

51. Letter to the President, 20 October 1959, Glennan Folder, Administration Series, Box 15, Whitman File, DDE Library.

52. Ibid.

53. Response from the President, 21 October 1959, Glennan Folder, Box 15, Administration Series, Whitman File, DDE Library.

54. W Brucker biographical information sourced from Bentley Historical Library, University of Michigan.

55. Executive Order Disposition Tables, 1960, National Archives, Washington, DC.

56. Glennan, *Edited Diary*, p. 23.

57. Ibid. Glennan was of the opinion that Adams, who resigned on October 7, 1958, shielded Eisenhower from "the nasty problems of organisational relationships"—also p. 23. However, Glennan's experience of Eisenhower prior to October 1958 would have been of a man recovering from a stroke.

58. Ibid.

59. Ann Whitman Diary Series, 1958–1960, DDE Library. Private Secretary Whitman's diaries give a daily schedule of Eisenhower's meetings. In the second half of the second term, her detailed noting of the comings and goings to the Oval Office indicate Persons and Press Secretary Hagerty were the most frequent visitors.

60. In the edited version of Glennan's Diary, the editor, Roger Launius, adds a footnote saying: "Contrary to a NASA priority, that these six [later increased to seven] astronauts be civilians, President Eisenhower directed that they come from the armed services' test pilot force." Glennan Diary, p. 20, footnote 23. But a later Glennan interview with JD Atkinson and JM Shafritz in 1982 used in *The Real Stuff: The History of NASA's Recruitment Program* (New York, 1985), Glennan stated that he, Gilruth, Dryden, and other senior NASA officials became convinced that they would not find enough test pilots with the requisite numbers of flying hours on fast jets and that service pilots, already accustomed to the rigors of high stress testing regimes, available at short notice and with fully-traceable service records, would be most appropriate. The interview states that Eisenhower cleared the request from Glennan "within five minutes."

61. There are many accounts of the Mercury selection process, primarily in astronaut memoirs and through the historical canon described in chapter one. T Wolfe, *The Right Stuff*, provides an entertaining though somewhat fictionalized account, while Kris Stoever (daughter of Scott Carpenter, one of the Mercury 7 astronauts) provides a well-written insight into the astronaut experience in *For Spacious Skies* (Harcourt, New York, 2002).

62. Richard Bissell was interviewed by William Burrows in Spring 1984, Burrows, *Deep Black*, p. 78.

63. Although Lovelace's appointment was made in July 1958, it was not formally made active until NASA opened for business on October 1, 1958. His committee was known as "The Lovelace Committee." http://history.nasa.gov/SP-4213/ch2.htm. Accessed August 30, 2014.

64. NASA News Conference: "Introducing the Mercury Astronaut Team," April 9, 1959 (NASA, Washington, DC, 1959) pp. 27–29.

65. Philip Taubman provides a pithy account of the Lovelace testing regime for the U-2 pilots in *Secret Empire*, pp. 158–63.

66. Killian, *Sputnik*, pp. 193–94.

67. Milton was president of Johns Hopkins University from 1956–1967 and again from 1971–1972.

68. The National Science Foundation timeline at http://www.nsf.gov/about/history/overview-50.jsp notes that NSF appropriation was almost doubled to $134 million for the twelve months beginning July 1, 1958. Funding for education more than tripled. Accessed August 30, 2014.

69. H York, *Race to Oblivion* (New York, 1970). York, who participated in PSAC's involvement in the test ban conferences, provides a compelling account of how a scientific study swiftly became a political minefield.

70. GB Kistiakowsky: Transcript of diary, 1959–1960, entry for September 21, 1959, Box 1, Kistiakowsky File, DDE Library.

71. JR Killian to DDE, May 27 1959, Box 4, U.S. President's Science Advisory Committee: Records, 1957–61, DDE Library.
72. Killian, *Sputnik*, p. 208.
73. Letter from DDE to JR Killian, July 16, 1959, Box 13, Killian Papers, MIT.
74. Letter from DDE to JR Killian, February 16, 1958, Box 13, Killian Papers, MIT.
75. Letter from DDE to JR Killian, April 20, 1959, Box 13, Killian Papers, MIT.
76. Letter from RM Nixon to JR Killian, June 1, 1959, Box 10, Killian Papers, MIT.
77. Wang, *In Sputnik's Shadow*, p. 40.
78. Kistiakowsky, *Scientist in the White House*, pp. 378–79. The Limited Nuclear Test Ban Treaty was finally signed on August 5, 1963. http://www.state.gov/t/isn/4797.htm. Accessed August 30, 2014.
79. By January 1960, the Soviets had opened a new Pacific Ocean target area, which pointed to a significantly increased range from their ICBMs.
80. Kistiakowsky, *Scientist in the White House*, p. 221.
81. Ibid., p. 225.
82. DDE: "Statement by the President on U.S. Achievements in Space," August 17, 1960, Online by Gerhard Peters and John T. Woolley, *The American Presidency Project*, http://www.presidency.ucsb.edu/ws/?pid=11911. Accessed September 1, 2014.
83. NASA Office of Program Planning and Evaluation, "The Long Range Plan of the National Aeronautics and Space Administration," Washington, DC, December 16, 1959, NASA Historical Collection (Washington, DC).
84. VM Zubok, *Khrushchev and the Berlin Crisis, 1958–1962*, The Cold War History Project, Working Paper 6, (Washington, DC, 1993).
85. P Stares, *The Militarization of Space: US Policy 1945–1984* (Ithaca, NY) 1985, p. 51.
86. AM Gallup and G Gallup, eds., *The Gallup Poll Cumulative Index: Public Opinion 1935–1997*, (Lanham, MD), 1999, entry 1672.
87. Ibid.
88. "Highlights in Space Exploration," Sections 60–92, Box 4, Space and Aeronautics, Rockets and Satellites, Pre-Presidential Papers, 1960 Campaign, Hamlin File, Richard M Nixon Presidential Library.
89. Ibid., p. 1.
90. Ibid.
91. Ibid.
92. CIA Director, Allen Dulles, wrote to the president, via General Goodpaster, in August 1960, to update him on the status of the Soviet ICBM threat, based on intelligence collected via U-2 overflights. The memorandum noted that the main Soviet strike threat came from its bomber force, but that "only a minimum long-range bomber production program is continuing in the Soviet Union." On ICBMs, the memorandum stated: "We have covered a number of the most highly suspect areas in the Soviet Union without having found a single launch site for operational missiles." The memorandum noted that there was no evidence of a crash program of ICBM production but also stated: "It is apparent that the Soviet ballistic missile program is a dynamic and growing program." "Statistics Relating to the U-2 Program," Memorandum from the Office of the Director, CIA to Brigadier General Andrew Goodpaster, 19 August, 1960, Missile Gap, Historical Collections Online, CIA, http://www.foia.cia.gov/sites/default/files/document_conversions/18/1960-08-19b.pdf. Accessed September 1, 2014. In February 1960, the CIA predicted the Soviet Union would have thirty-six ICBMs operational by the end of the year. The reality was that two missiles were actually operational—see *Soviet Capabilities for Strategic Attack through 1964, National Intelligence Estimate* Source NIE 11-8-59, 9, February 1960, Missile Gap, Historical Collections Online, CIA, http://www.foia.cia.gov/sites/default/files/document_conversions/18/1960-02-09a.pdf. Accessed September 1, 2014. Meanwhile, the United States had deployed its first ICBMs on October 31, 1959, at Vandenberg Air Force Base in California, and had six missiles in place by the end of the year.
93. "Statement on US Achievements in Space" released at the President's News Conference, Washington, DC, August 17, 1960, http:presidency.ucsb.edu/ws/index.

php?pid+11910. Accessed September 1, 2014.

94. Glennan Diary Transcript, October 11, 1960.

95. Introduction to "Report of the Ad Hoc Panel on Man in Space," December 16, 1960, NASA Historical Reference Collection, History Office, NASA Headquarters, (Washington, DC).

96. Ibid., Section 2—The Man in Space Program.

97. Ibid.

98. Ibid.

99. The Nova class rocket, as envisaged by NASA, was planned to have eight first stage engines delivering twelve million lbs of thrust. A second stage of four clustered engines would deliver a further four million lbs' thrust, and a final stage would add a further 200,000 lbs of thrust. The only use of Nova, which would be capable of lifting a 160,000 lb payload would be for a direct ascent manned moon mission. Murray and Bly Cox, *Apollo—The Race to the Moon,* pp. 109–10.

100. *Man in Space,* Section 4—Relation between Manned and Unmanned Space Exploration.

101. Ibid., Section 6—Conclusions.

102. Glennan Diary, December 20, 1960.

103. RC Seamans Jr., *Project Apollo—The Tough Decisions,* NASA Monographs in Aerospace History, no. 37, (Washington, DC, 2005).

104. Glennan Diary: January 3, 1961.

105. US Congress, Senate Committee on Appropriations, "Independent Offices Appropriations, 1962," 87th Congress, 1st session, 1961, p. 653.

106. Task Force Reports: Space, Box 1072, Transition Files, JFK Pre-Presidential Papers, John F Kennedy Presidential Library.

107. Kistiakowsky, diary entry for July 12, 1960.

108. Kistiakowsky diary entry for August 22, 1960.

109. Ibid.

110. Kistiakowsky diary entry, September 27, 1960.

111. Ibid.

112. Task Force Report: Space, p. 24.

113. Ibid., p. 19.

114. Ibid.

115. Ibid.

116. Ibid.

117. Ibid., pp. 19–20.

118. Janis, *Groupthink,* p. 35.

119. Killian, *Sputnik,* p. 241.

120. Transcript, Donald F. Hornig Oral History Interview I, 12/4/68, by David G. McComb, http://www.lbjlib.utexas.edu/johnson/archives.hom/oralhistory.hom/Hornig-D/hornig1.PDF, LBJ Library, p14. Accessed August 30, 2014.

121. Ibid., p. 13.

122. M Mahony, "Having the President's Ear" in *Technology Review* (MIT, Cambridge, MA), September/October 2008, p. 112.

123. K Collier, "Eisenhower and Congress: The Autopilot Presidency," *Presidential Studies Quarterly,* Vol 4, No. 2, (New York, Spring 1994), pp. 309–25.

124. Kistiakowsky, *Scientist in the White House,* p. lv1.

125. Charles Duke in conversation with the author, Birmingham Hilton Metropole Hotel, England, 24 April 2009.

SIX

Final Thoughts

Ike Revisited on Space

For too long, the popular narrative has observed Eisenhower's moves to develop a first space policy for the United States merely as a reaction to Sputnik. This has been part of an overall view of Eisenhower's second term as the epitome of "lame-duck" presidency with little achieved and meandering to a close. That is wrong in both cases. Eisenhower's space policy owed much more to decisions taken following the work of the TCP in 1955 than to any Sputnik effect. While he was determined to avoid a "space race," he did put in place a coherent space policy. What emerges is a picture of rational actions backed by sound judgment on the part of Eisenhower. It shows a logical progression. From his inheritance of competing armed forces rocket programs, through his appreciation of "missile power" following the 1954 Castle series of H-bomb tests to the opportunities presented by the International Geophysical Year's satellite development, via the national security need for global reconnaissance, he set in motion a process that enabled the creation of an agency that could deliver an effective space program. It showed how that program had the tools and impetus to open space in its widest sense to the American spirit of exploration, while simultaneously advancing the national security agenda. Imperative to this, it charted how Eisenhower played his strategic, "long game," marshalling the forces at his disposal to reach sound outcomes without causing national or international alarm—and all the while, working towards a balanced budget. We have seen the president's role in key decision making, balanced against the influence of a small number of advisers both on the "big ideas" for Eisenhower's space policy and, indeed, the series of small decisions that cemented the fine detail of the policy and how that policy played out once NASA was operational.

Alongside this, this book has analyzed the contribution of previously under-recognized players in policy development: Killian, Hagerty, Purcell, and Bissell among others. It has also re-evaluated the contribution of Keith Glennan while assessing the contribution made by Washington politics' original "Mr. Space," Lyndon Johnson. In each case, *Eisenhower at the Dawn of the Space Age* has explained why these figures were chosen by Eisenhower above others—in political, military, and science circles—when there may have been other voices to the fore that may have seemed more obvious choices to advance the executive agenda. It showed how Eisenhower was able to use each of them to advance his strategy—whether that was helping him to refine that strategy into something operationally-achievable or, more often, to implement it.

This adds evidence to David Nichols's claim, made in New York in 2013 and extended in Andy Polsky's edited collection, *The Eisenhower Presidency, Lessons for the Twenty First Century*[1] that research on Eisenhower is "in mid-stream."[2] Nichols noted that historians were finally moving away from the popular perception of the first generation of scholars to explore the Eisenhower presidency: "Ike is not the fictional war hero who came home, assumed the presidency and morphed into a bumbling, smiling, over-the-hill grandfather more interested in playing golf than being a President. . . . We still have colleagues who resist the conclusion that Ike was a demonstrably competent, even extraordinary President. No more. Our obligation now is to write the history of the Eisenhower presidency with rigorous accuracy, not political pay-off."

This book also adds to the ongoing wider revision of Eisenhower. In addressing the role Sputnik played, it demonstrates his calm handling of an incendiary situation, where his consistency did not descend into dogmatism. President Eisenhower played a long-game strategic hand in developing the United States' first space policy. He rejected spectaculars, but encouraged the best scientific brains of the nation to enable his administration to pursue a parallel track: outside the partisan political process and adjacent to, but never fully within, the military-industrial complex. That parallel path may not have been delivered with the rhetorical flourish of his successors, but it was a formidable achievement from a formidable president who used the best brains to achieve a powerful result.

By examining the evidence, it becomes clear that Eisenhower was a more complex president than often suggested. He was not the amiable, golf-club swinging, dissembler as caricatured by the first wave of historians to take interest in him. But he did present such a face to the public when it suited him. Nor was he the all-seeing, all-knowing, behind-the-scenes manipulator as portrayed by Greenstein's "straw man" political science model and the more extreme revisionists. What emerges is a president who operated with a degree of confidence rare even in the White House. He was well-read, on top of his brief, and driven by an overriding desire for small government, peaceful foreign relations, and economic growth fueled by business success. Eisenhower was not always right. However, he had a sufficient streak of vanity to believe that he *should* be. He favored a military structure and approach to running executive government, but aimed to deliver this through modern business methods—and that meant installing senior business figures in key governmental roles. He was stubborn, but not unbending; a cold rationalist, but one who maintained warm relations with those who met his expectations in delivering solutions. He had a massive sense of duty to do the right thing for the nation, and believed he understood better than anyone else what that "right thing" really was. He was an ardent anti-communist, but not driven by the insidious fear that riddled the nation in the wake of McCarthy. Most of all, he was a calm strategist, prepared to meet the external threats to the United States with just enough show of force to deter that threat from ever becoming a reality.

In recapping the key findings of this book, it is worth reminding readers that Eisenhower had all the elements for an effective space policy in place as early as 1955, two years before the Sputnik satellites. Using the special study process he favored, led by Jim Killian who was to prove such a useful presiding officer in the development of space-related policy, he was able to define effective means both to discern the real threat the Soviet Union posed to the United States, and to prioritize the missile development that was to be at the heart of his New Look defense policy. Building from this, we saw how the implementation of the post-TCP policy delivered high quality intelligence gathered by the CIA on Soviet missiles and bomber deployment via numerous U-2 over-flights as early as the summer of 1956 that demonstrated clearly that the Soviets did not pose a strategic nuclear threat. Based on this evidence, Eisenhower had no need to develop new strategies in response to Sputnik and therefore did not do so. He already had a plan in place both to develop and deploy US ICBMs and IRBMs and, quite separately, to enable a non-military program for the exploration of outer space and use of satellite technology for beneficial activities such as communication, navigation, and meteorology. He was well-attuned too to the benefits of operations in the "black" arena where the programs intersected, and enabled the development of an effective secret reconnaissance system, *CORONA*, hidden in plain sight as *Discoverer*. Eisenhower continued along a parallel track, refining his military space program to focus it on the Air Force, and creating the civilianized NASA as a logical next step from the initial scientific efforts in space first detailed in 1955 during the planning for the 1957–1958 International Geophysical Year.

The hinge chapters in this work demonstrate that Eisenhower was effective through drawing on the expertise of actors from all political spheres, whether elected politicians such as Johnson, political insiders, such as Hagerty, and those on the edge or outside the traditions of Washington politics such as Killian, York, Johnston, and Dembling who were crucial in aiding his understanding of what was both possible and desirable in space and how it could be both organized and enabled. In essence, the "Helping Hands" enabled his success.

As the previous chapter in particular showed, Eisenhower was active and interventionist when he needed to be, but also had high expectations that his appointees would deliver his required strategy with little need for intervention. His priorities in space were always led by the abiding priority he gave to national security, and supported by a core tenet of fiscal conservatism. First and foremost, rockets were intended to enable the United States' defense as a fundamental element of the New Look defense strategy. But this was a president who admired scientists and their achievements and had faith in the process of moving the nation forward through scientific advance. Thus there was a place for scientific discovery, not least, in the newly-reached realms of outer space. But this discovery would be an incremental effort, largely conducted by unmanned probes, which would benefit from the same family of technological advances that would spawn communications, reconnaissance, navigation, and meteorology satellites. Such development would be managed by the best and brightest scientific and engineering minds that would create prestige for the United States not through the "smoke and mirrors" of Khrushchev-style propaganda stunts, but through tangible scientific discoveries that would advance the knowledge of the nation. Eisenhower simply did not believe in stunts or spectaculars.

Sputnik was just such a stunt, and to react would have plunged a president committed to waging peace into an arena of waging a proxy Cold War battle in space. Eisenhower, the rather cold, slightly insular, pragmatic general saw no virtue in such a battle. While a short-lived furor lit up sections of the media and offered a platform for political beasts circling the next presidential election campaign, the actual impact of Sputnik on the United States was shorter-lived and considerably less deep than conventional interpretations by both scholars of Eisenhower and writers on the space program have conceded.

Eisenhower set a clear direction for US policy in space. He gave the country NASA; the Mercury man-in-space program; a large satellite lead over the Soviets, interplanetary probes, and the Saturn super-booster. He organized a workable split between the military satellite programs, ICBM development and deployment, and civilianized activities, and he managed to do so while balancing the national budget. In order to achieve this, he drew in a circle of remarkable "presiding officers" from Killian through Kistiakowsky, Glennan, Purcell, Land, and Bissell while being supported by unexpected advisers, most especially Hagerty. His actions were not perfect, but understandable. He should have been more decisive between 1955 and 1958 in directing ICBM/IRBM development, and he should have been more visible and reassuring in the immediate aftermath of Sputnik. Here, his failing was not one of reaction, but of failing to grab the media high ground. Equally, he should have stayed more closely in touch with NASA's planning in 1959/60 as the new agency began to stretch and flex away from the president's initial intent. But the fundamental finding of this study is that Eisenhower's actions on space from 1954 to the end of his presidency are undervalued. The building blocks he put in space deserved much more than a space program that reached the moon, but has been falling earthwards ever since.

It is time to reconsider the chronology of the so-called "space race" where the norm is to bound it by Sputnik as a warning shot, to regard Kennedy's "Second State of the Union" in 1961 as the call to action, and to view Armstrong and Aldrin's moon landing as the satisfactory conclusion. It is evident that historians should look considerably earlier than Sputnik for the genesis of the United States' participation in space. It is more appropriate to consider the decision to launch an earth-satellite as America's contribution to the IGY as the true catalyst, a decision derived from the work of the TCP that also led to work on the U-2 and CORONA reconnaissance programs while leading to top prioritization for the development of ICBMs and IRBMs. It is time to reconsider Eisenhower's presidency; his pragmatic contribution to reconnaissance technology development and definitely time to reappraise his position in US space history.

It is time too to reconsider the orthodox view of Eisenhower as a "lame-duck" president from 1957 onwards, powerless to confront a hostile Congress. From the outset, national security was foremost among Eisenhower's priorities. Yet that was balanced by the fundamental principle of achieving that security without spending a cent more than was necessary. In practice, this was not always achieved. But Eisenhower's prime means to achieve it was by keeping control of the issue. To do this, Eisenhower created new bureaucracies or changed the purpose of bureaucratic functions to achieve his ends. The TCP gave him the control he needed to define and enable the core elements of his New Look defense policy. Developing the U-2, A-12, and CORONA through the CIA enabled him to circumvent the armed services to deliver these projects on time and at a far lower financial cost to the country while maintaining executive control. Finally, in entrusting the drafting of an executive-sponsored bill for the creation of NASA to an expert team led by Killian, he ensured that the legislative act that emerged delivered an agency under presidential rather than congressional control.

There is a belief that if we hear a story often enough it becomes true. The popular reading of Sputnik as a prelude to the race to the moon is not true. The challenging of a "lame-duck" presidency must continue. It is important that in recognizing Eisenhower's parallel track that scholars cut through the myth and reframe the US space narrative. It is time we gave due credit to Dwight D Eisenhower.

NOTES

1. A. Polsky, "Shifting Currents" in *The Eisenhower Presidency* (Lanham, MD, 2015), p. 227.
2. D Nichols, closing remarks to "Ike Reconsidered" conference, March 8, 2013, Roosevelt House, Hunter College, CUNY.

Bibliography

PRIMARY SOURCES

Unpublished

Dwight D Eisenhower Presidential Library, Abilene, Kansas, USA—consulted collections:

THE COLD WAR IN THE EARLY FIFTIES:
"Open Skies" and the 1955 Geneva Summit

Series consulted:
Dulles, John Foster: Papers
Eisenhower, Dwight D: Papers as President
Ann Whitman Diary Series
Cabinet Series
DDE Diary Series
Press Conference Series
General File
Hagerty, James C.: Papers

CORONA, SAMOS, Anti-satellite Weapons Program
Series consulted:
Bissell, Richard M., Jr.: Papers
Eisenhower, Dwight D.: Papers as President of the U.S., 1953–61 (Ann Whitman File)
Administration Series
ACW Diary Series
Cabinet Series
DDE Diary Series
NSC Series
Eisenhower, Dwight D.: Records as President of the U.S. (White House Central Files), 1952–1961
Confidential File
Official File
Emme, Eugene: Papers
Gray, Gordon: Papers
U.S. President's Science Advisory Committee: Records, 1957–1961
White House Office, National Security Council Staff
Executive Secretary's Subject File
White House Office, Office of the Special Assistant for National Security Affairs: Records,
NSC Series, Policy Papers Subseries
NSC Series, Subject Subseries

Intelligence:
National Security Council Intelligence Briefings
The Report of the Technological Capabilities Panel (Killian Report): Meeting the Threat of Surprise Attack (February 1955)
Cold War Reconnaissance—Airborne Peripheral Reconnaissance missions
Cold War Reconnaissance—U-2 Over-flights
Cold War Reconnaissance: Satellite Reconnaissance

ORAL HISTORIES:
Richard Bissell (OH–168, OH–382)

INTERNATIONAL GEOPHYSICAL YEAR
Series consulted:
Eisenhower, Dwight D.: Papers as President (Ann Whitman File)
Speech Series:
Eisenhower, Dwight D.: Records as President (White House Central Files)
Official File:
General File:
U.S. President's Committee on Scientists and Engineers: Records, 1956–1958
White House Office, National Security Council Staff: Papers, 1948–1961

ORGANIZATION AND EARLY HISTORY OF NASA, 1957–1961
Series consulted:
Dulles, John Foster: Papers
Gerard Smith Series
JFD Chronological Series
Subject Series
White House Memoranda Series
Eisenhower, Dwight D.: Papers as President
Ann Whitman Diary Series
Cabinet Series
DDE Diary Series
Dulles-Herter Series
Legislative Meetings Series
Legislative Leaders Meetings 1957 (3) 5/14/57 [missiles]
NSC Series
Press Conference Series
Speech Series
Eisenhower, Dwight D.: Post-Presidential Papers
Eisenhower, Dwight D.: Records as President
Confidential File
National Aeronautics and Space Council
General File
Administrator
Official File
Emergency Preparedness, Office of: Printed Material, 1953–1961
Ford, Vincent T.: Manuscript Material
Glennan, T. Keith: Diary
Hagerty, James C.: Papers
Harlow, Bryce: Records
Herter, Christian: Papers
Hess, Stephen H.: Papers
Larson, Arthur: Papers
Larson, Arthur and Malcolm C. Moos: Papers
McElroy, Neil H.: Papers

National Aeronautics and Space Administration: Documents relating to the Space Program
Persons, Wilton B.: Records
Quarles, Donald A.: Papers
Seaton, Fred: Papers
U.S. President's Advisory Committee on Government Organization (Rockefeller Committee): Records
U.S. President's Science Advisory Committee: Records
White House Office, Office of Budget and Accounting Records
White House Office, Office of the Special Assistant for Executive Appointments
White House Office, Office of the Special Assistant for National Security Affairs
White House Office, Office of the Special Assistant for Science and Technology: Records
White House Office, Office of the Staff Secretary: Records

Oral Histories:
Gavin, James M. (OH–184) [satellites, missiles]
Hagerty, James C. (OH–91) [satellite anecdote]
Hauge, Gabriel (OH–190) [Sputnik]
Killian, James R., Jr. (OH–402) [NASA, Sputnik, space, missiles, satellites]
Kistiakowsky, George (OH–412) [NASA, space, satellites]
Knowland, William F. (OH–333) [Sputnik]
McCone, John A. (OH–201) [Saturn rocket]
McCone, John A. (OH–396) [space program]
McElroy, Neil (OH–26) [satellites, missiles, Sputnik, rockets]
Rockefeller, Nelson (OH–231) [space]

U-2, U-2 INCIDENT AND PARIS SUMMIT OF 1960
Series consulted:
Bissell, Richard M., Jr.: Papers
Eisenhower, Dwight D.: Papers as President (Ann Whitman File)
Administration Series:
ACW Diary Series
DDE Diary Series
Dulles-Herter Series
NSC Series
Press Conferences Series
Speech Series

Oral Histories:
Richard Bissell (OH–168)
Richard Bissell (OH–382)
Charles Bohlen (OH–136)
Dwight D. Eisenhower (OH–14)
Andrew Goodpaster (OH–378)
James Hagerty (OH–91)

NASA HQ HISTORICAL REFERENCE COLLECTION, Washington, DC, USA
Origins of NASA collection
Media Collection

MIT LIBRARY, Cambridge, Massachusetts, USA
James R Killian Papers

NATIONAL LEGISLATIVE ARCHIVE, Washington, DC, USA

Congressional Record

HARRY S TRUMAN PRESIDENTIAL LIBRARY, Independence, Missouri, USA
K.T. Keller Papers, Director, Office of Guided Missiles, Office of the Secretary of Defense, 1950–1953.

JOHN F KENNEDY PRESIDENTIAL LIBRARY, Columbia Point, Boston, Massachusetts, USA
JFK Pre-Presidential Papers, Transition Files

RICHARD M NIXON PRESIDENTIAL LIBRARY, Yorba Linda, California, USA
RMN Pre Presidential Papers, 1960 Campaign Papers, Hamlin file

NATIONAL SECURITY ARCHIVE/CREST ARCHIVE, Washington, DC, and College Park Maryland, USA
Deterrence and Survival in the Nuclear Age, Security Resources Panel of the Science Advisory Committee Report to the President, November 7, 1957, http://www2.gwu.edu/~nsarchiv/NSAEBB/NSAEBB139/nitze02.pdf

Published

Documentary Collections:
Foreign Relations Series, *Foreign Relations of the United States,* 1952–1954, National Security Affairs, Volume 2, part 1
Foreign Relations Series, *Foreign Relations of the United States,* 1955–1957; Volume 11, United Nations and General International matters

Technical Reports:
AJ Wohlstetter, FS Hoffmann and HS Rowen, *Project RAND, Protecting US Power to Strike Back in the 1950s and 1960s* September 1, 1956 (Santa Monica, CA)

Newspaper/News Magazine Primary source material:
Daily Express (London) sourced from British Library Newspaper Collection
Daily Herald (London) sourced from British Library Newspaper Collection
Life Magazine (Chicago), sourced from British Library
Paris-Journal (Paris, France) sourced from the British Library Newspaper Collection
Washington Post, sourced from British Library Newspaper Collection
LA Times, sourced from British Library Newspaper collection
Chicago News Tribune, sourced online from www.chicagotribune.com
Atlanta Constitution, sourced from British Library Newspaper Collection
Time Magazine (New York), sourced online from www.time.com
The New Republic (New York), sourced from Megan Guernsey, archivist at tnr.com
Associated Press, Scripps International and United International Press syndicated material from newspapers.com published in:
Athens Limestone Democrat (Athens, Alabama)
The Atlanta Constitution (Atlanta, Georgia)
Baltimore Sun (Baltimore, Maryland)
The Courier News (Blytheville, Arkansas)
The Corsicana Daily Sun (Corsicana, Texas)
El Paso Herald Post (El Paso, Texas)
Idaho State Journal (Pocatello, Idaho)
Kansas City Star (Kansas City, Missouri)
Louisville Courier-Journal (Louisville, Kentucky)
The Mansfield News-Journal (Mansfield, Ohio)

Miami Daily News-Record (Miami, Oklahoma)
The Milwaukee Sentinel (Milwaukee, Wisconsin)
Pasadena Independent (Pasadena, California)
Raleigh Register (Raleigh, North Carolina)
The Salina Journal (Salina, Kansas)
Schenectady Gazette (Schenectady, NY)
The Sun-Telegram (San Bernardino, California)
Tucson Daily Citizen (Tucson, Arizona)

Oral Histories:
Legislative Origins of the National Aeronautics and Space act of 1958, Proceedings of Oral History Workshop, moderated by JM Logsdon and published as *Monographs in Aerospace History Number 8* (NASA History Office, Washington, DC, 1998)
Richard M Bissell Jr., 9/7/1971, by Theodore A. Wilson and Richard D, McKinzie, Internet Copy, Harry S Truman Presidential Library
Lucius Clay, Oral History OH–56, Dwight D Eisenhower Presidential Library via National Archives
Donald F. Hornig, Oral History Interview I, 12/4/68, by David G. McComb, Internet Copy, Lyndon Baines Johnson Presidential Library.
Richard M Nixon, 15/5/91, Eisenhower Legacy Collection, produced by George Colburn, American History TV, C-Span 3

Memoir:
RM Bissell, *Reflections of a Cold Warrior* (New Haven, CT, 1996)
DD Eisenhower, *The White House Years: Mandate for Change* (London, 1963)
DD Eisenhower, *The White House Years: Waging Peace* (New York, 1965)
DD Eisenhower, *At Ease* (New York, 1967)
VT Ford, *Twenty-Four Minutes to Checkmate,* unpublished manuscript Box 2, Volume 3, part 3, DDE Library
TK Glennan, *The Birth of NASA* (Washington, 1993)
S Khrushchev's translation of *Khrushchev Remembers* (University Park PA, 2004)
J Killian, *Sputnik, Scientists and Eisenhower* (Cambridge, MA) 1977
G Kistiakowski, *A Scientist at the White House* (Cambridge MA, 1976)
LB Johnson, *The Vantage Point* (New York, 1971)
R Nixon, *Six Crises* (New York, 1962)
HF York, *Making Weapons, Talking Peace* (New York, 1997)

Diaries:
RH Ferrell *The Diary of James C Hagerty* (Bloomington, IN, 1983)
RH Ferrell *The Eisenhower Diaries* (New York, 1981)
DD Holt and JW Layersaf eds, *Eisenhower: The Pre-War Diaries and Selected Papers 1905–1941* (Baltimore, 1998)

SECONDARY SOURCES

C Allen, *Eisenhower and the Mass Media* (Chapel Hill, NC, 1993)
M Allen, *Live from the Moon—Film, Television and the Space Race* (London, 2009)
GT Allison, *Essence of Decision* (Glenview, IL, 1971)
S Ambrose, *The Supreme Commander* (New York, 1970)
S Ambrose, *Eisenhower, Soldier and President* (New York, 2003)
S Ambrose, *Nixon* (New York, 1987)
S Ambrose and R Immerman, *Ike's Spies: Eisenhower and the Espionage Establishment* (New York), 1981
G Baldasty, *EW Scripps and the Business of Newspapers* (Urbana, IL, 1999)

J Barbree, *Live from Cape Canaveral* (Washington, DC, 2008)
JD Barber, *The Presidential Character: Predicting Performance in the White House*, 4th ed. (New York, 1992)
JL Baughman, *Henry R Luce and the Rise of the American News Media* (New York, 1987)
J.L. Baughman, *Same Time, Same Station* (Baltimore, 2007)
TA Becnel, *Senator Allen Ellender of Louisiana* (Baton Rouge, 1995)
P Bizony, *The Man Who Ran the Moon* (Cambridge, 2006)
H Brogan, *The Penguin History of the USA* (London, 2009 edition)
JL Bromberg, *NASA and the Space Industry* (Baltimore, 1999)
CG Brooks, J.M. Grimwood, L Swenson, *Chariots for Apollo: The NASA History of Manned Lunar Spacecraft to 1969* (Mineola, NY, 2009 reprint of 1979 edition)
M Brzezinski, *Red Moon Rising* (London, 2007)
WE Burrows, *This New Ocean: History of the First Space Age* (New York, 1999)
WE Burrows, *Deep Black* (London, 1998)
M. Byrnes, *Politics and Space—Image Making by NASA* (Westport, CT, 1994)
D Cadbury, *Space Race* (London, 2006)
RA Caro, *Lyndon Johnson Master of the Senate* (New York, 2003)
S Carpenter and K Stoever, *For Spacious Skies* (Orlando, FL, 2002)
E. Cernan and D Davis, *The Last Man on the Moon* (New York, 2000)
A Chaikin, *A Man on the Moon* (London, 1998)
DG Coleman and JM Siracusa, *Real-World Nuclear Deterrence* (Westport, CT, 2008)
K Collier, Eisenhower and Congress, "The Autopilot Presidency" *Presidential Studies Quarterly*, Vol 4, No. 2 (New York, Spring 1994)
M Collins, *Carrying the Fire* (New York, 2001)
W Cunningham, *The All-American Boys* (New York, 2004)
W Cunningham, *Spaceflight* Vol 5, no. 7 (London, July 2009)
R Dallek, *Johnson: Portrait of a President* (London, 2004)
R Dallek, *Kennedy, An Unfinished Life* (London, 2003)
PHJ Davies, *Intelligence and Government in Britain and the United States, Volume 1: Evolution of the US Intelligence Community* (Santa Barbara, CA, Denver, CO and Oxford, 2012)
G DeGroot, *Dark Side of the Moon* (London, 2008)
P Dickson, *Sputnik: The Shock of the Century* (New York, 2001)
R Divine, *The Sputnik Challenge* (Oxford, 1993)
R Divine, *Eisenhower and the Cold War* (Oxford, 1981)
S Dockrill, *Eisenhower's New Look national security policy 1953–1961* (London, 1996)
JH Doolittle with CV Glines, *I Could Never Be So Lucky Again* (New York, 1991)
J Doran and P Bizony, *Starman* (London, 1998)
BR Duchin, "The Most Spectacular Legislative Battle of the Year: President Eisenhower and the 1958 Reorganization of the Department of Defense," *Presidential Studies Quarterly*, Vol. 24, No 2, Eisenhower and Governance (1994)
GK Ericksen, *Women Entrepreneurs Only: 12 Women Entrepreneurs Tell the Stories of Their Success* (New York, 1999)
W Ewald, *Eisenhower the President* (Englewood Cliffs, NJ, 1981)
J Frank, *Ike and Dick* (New York, 2013)
N Frazier, *William Randolph Hearst, Modern Media Tycoon* (Woodbridge, CT, 2001)
T Furniss, *One Small Step* (Yeovil, 1989)
JL Gaddis, *The Cold War* (London, 2005)
JL Gaddis, *Strategies of Containment* (Oxford, 1982)
AM Gallup and G Gallup, eds *The Gallup Poll Cumulative Index: Public Opinion 1935–1997* (Lanham, MD, 1999)
F Greenstein, *The Hidden Hand Presidency* (Baltimore, 1982)
F Greenstein, *The Reagan Presidency: An Early Assessment* (Baltimore, 1983)
R Griffith, "Dwight D Eisenhower and the Corporate Commonwealth," *American Historical Review* 87 (February 1982,)
JM Grimwood, *Project Mercury: A Chronology* (NASA SP-4001, Washington, DC 1963)

E Goldstein, GC La Rosa, and DS Schuman, *Present at the Creation* (Washington, DC, 2008)
A Griffith *The National Aeronautics and Space Act* (Washington, DC, 1962)
MA Guhin, *John Foster Dulles, A Statesman And His Times* (New York and London, 1972)
K Hall Jamieson, *Packaging the Presidency* (Oxford, 1984)
R Hall and D Shayler, *The Rocket Men* (Chichester, 2001)
JR Hansen, *Spaceflight Revolution: NASA Langley Research Centre from Sputnik to Apollo* (Washington, DC, 1995)
JR Hansen, *First Man* (New York, 2005)
V Hardesty and G Eisman, *Epic Rivalry* (Washington, DC, 2007)
J Harford, *Korolev* (London, 1997)
T Heppenheimer, *Countdown* (London, 1997)
MJ Hogan, *A Cross of Iron, Harry S. Truman and the Origins of the National Security State, 1945–54* (Cambridge, 1998)
M Holland, *Eisenhower Between the Wars* (Westport, CT, 2001)
E Hughes, *The Ordeal of Power* (New York, 1962)
KH Jamieson, *Packaging the Presidency* (Oxford, 1984)
IL Janis, *Groupthink*, 2nd edition (Boston, MA, 1972)
P Johnson, *Eisenhower: A Life* (New York, 2015)
J Johnson-Freese, *Space as a Strategic Ssset* (New York, 2007)
SN Kalic, *US Presidents and the Militarization of Space, 1946–1967* (College Station, TX, 2012)
A Kasten Nelson, "The 'Top of Policy Hill': President Eisenhower and the National Security Council," *Diplomatic History*, Volume 7, issue 4, 1983
J Kauffman, *Selling Outer Space—Kennedy, the Media and Funding for Project Apollo, 1961–1963* (Tuscaloosa, AL, 1994)
K Keltner, *Live at the (Project) Apollo: How CBS Reported the Moon Story* (Nashville, TN, 2005)
F Kohler and D Harvey, "The International Significance of The Lunar Landing, *Journal of Inter-American studies and World Affairs* Vol 12, No. 1 (Jan 1970)
G Kranz, *Failure Is Not an Option* (Berkley CA, 2000)
R Launius and H McCurdy, *Spaceflight and the Myth of Presidential Leadership* (Champaign, IL, 1997)
R Launius, *Apollo a retrospective analysis* (Washington, DC, 2004)
R Launius, "An Unintended Consequence of the IGY: Eisenhower, Sputnik, the Founding of NASA," *Acta Astronautica* 67 (Oxford, 2010),
R Launius, "Sphere of Influence: The Sputnik Crisis and the Master Narrative," *Quest, the History of Spaceflight Quarterly* Vol. 14, No. 4 (2007)
R Launius, "Kennedy's Space Policy Reconsidered: A Post-Cold War Perspective," *Air Power History*, Vol 50, No. 4 (Winter 2003)
W LeFaber, *America, Russia and the Cold War, 1945–1996*, 8th edition (New York, 1997)
W Leuchtenburg ed, "First Republican Interlude: Dwight D Eisenhower," *In the Shadow of FDR* (Ithaca, NY, 1983)
D Levy, *Lyndon B Johnson (Presidential Leaders)* (Minneapolis, 2003),
W Lippmann, *Public Opinion* (New York, 2010)
JM Logsdon, *Exploring the Unknown, Selected Documents in the History of the US Civil Space Program* (Washington, DC, 1995)
JM Logsdon, *John F Kennedy and the Race to the Moon* (New York, 2010)
JC Lonnquest and DF Winkler, *To Defend and Deter, The Legacy of the US Cold War Missile Program* (Washington, DC, 1996)
D MacIsaac, "Eisenhower, A Reputation in Transition," *Air University Review* (September–October 1982)
M Mahony, "Having the President's Ear," *Technology Review* (MIT, Cambridge, MA, September/October 2008)
N Mailer, *A Fire on the Moon* (New York, 1970)

H Makemson, *Media, NASA and America's Quest for the Moon* (New York, 2009)
M Mayer, *The Eisenhower Presidency and the 1950s* (New York, 1998)
RA McDonald, *CORONA between the Sun and the Earth* (Washington, DC, 2007)
W McDougall, *The Heavens and the Earth—A Political History of the Space Age* (Baltimore, 1997)
K McQuaid, "Sputnik Reconsidered: Image and Reality in the Early Space Age," *Canadian Revue of American Studies*, Vol 37, no. 3 (2007)
Mercury Astronauts, *We Seven* (New York, 1959)
R Melanson and D Mayers, *Re-evaluating Eisenhower: American Foreign Policy in the 1950s* (Urbana, IL, 1987)
Y Mieczkowski, *Eisenhower's Sputnik Moment* (Ithaca, NY, 2013)
C Murray and C Bly Cox, *Apollo, The Race to the Moon* (New York, 1989)
P Nash, *The Other Missiles of October* (Chapel Hill, NC, 1997)
M Neufeld, *Von Braun: Dreamer of Space, Engineer of War* (Washington, DC, 2008)
RE Neustadt, *Presidential Power* (New York, 1960)
J Newton, *Eisenhower, The White House Years* (New York, 2012)
DA Nichols, *A Matter of Justice* (New York, 2007
DA Nichols, *Eisenhower 1956* (New York, 2011)
B O'Leary, *The making of an ex-astronaut* (Windsor, CT, 1971)
BR Page, "The Creation of NASA," *The Journal of the British Interplanetary Society*, Vol. 32 (London, 1979)
C Peebles, *High Frontier, The USAF and the Military Space Program* (Washington, DC, 1997)
C Peoples, "Sputnik and Skill Thinking Revisited," *Cold War History*, Vol 8, No 1 (Feb 2008)
A Polsky, ed., *The Eisenhower Presidency—Lessons for the Twenty First Century* (Lanham, MD, 2015)
S Rabe, *Eisenhower and Latin America* (Chapel Hill, NC, 1988)
D Reynolds, *America, Empire of Liberty* (London, 2009)
E Richardson, *The Presidency of Dwight D Eisenhower* (Lawrence, KS, 1979)
R Seamans, *Project Apollo, the Tough Decisions* (Washington, DC, 2005)
J Schefter, *The Race* (New York, 1999)
AM Schlesinger, *The Imperial Presidency* (New York, 2004)
D Scott and A Leonov, *Two Sides of the Moon* (Simon and Schuster, 2004)
N Sheehan, *A Fiery Peace in a Cold War* (New York, 2010)
A Shepard and D Slayton, *Moonshot* (London, 1994)
A Siddiqi, "American Space History: Legacies, Questions and Opportunities for Future Research," *Critical Issues in the History of Spaceflight* ed. by S Dick and R Launius (NASA, Washington, DC, 2006)
A Siddiqi, *The Soviet Space Race with Apollo* (Gainesville, FL, 2003)
JE Smith, *Eisenhower in War and Peace* (New York, 2012)
Smithsonian Air & Space panel discussion, "In Their Own Words" (November 2000). Key participants, Julian Scheer (Head of NASA Public Affairs during Apollo) and Walter Cronkite
TC Sorensen, *Decision Making in the White House* (New York, 1963)
TC Sorensen, ed., *Let the Word Go Forth* (New York, 1988)
P Stares, *The Militarization of Space: US Policy 1945–1984* (Ithaca, NY, 1985)
G Swanson, *Before This Decade is Out* (Gainesville, FL, 2002)
P Taubman, *Secret Empire* (New York, 2003)
E Thomas, *Ike's Bluff* (New York, 2012)
L Thomas and E Jablonski, *Doolittle, A Biography* (New York, 1976)
G Tindall and D Shi, *America, A Narrative History*, 7th edition, volume 2 (New York, 2007)
R Turnill, *The Moon landings, An Eyewitness Account* (Cambridge, 2003)
W Tompson, *Khrushchev, A Political Life* (Basingstoke, 1977)

J Van Nimmen, L Bruno & R Rosholt, *NASA Historical Data Book, 1958–68* (Washington, DC, 1976)
Z Wang, *In Sputnik's Shadow* (New Brunswick, NJ, 2008)
R Ward, *Dr Space, the Life of Werner Von Braun* (Annapolis, MD, 2005)
DE Welzenbach, "Science and Technology: Origins of a Directorate in Studies," *Intelligence* Vol. 30, No. 2 (1986)
B Wiesen Cook, *The Declassified Eisenhower* (New York, 1984)
M Winn, *The Plug-in Drug* (London, 2002)
T Wolfe, *The Right Stuff* (New York, 2005)
C Wright Mills, *The Power Elite* (Oxford, 1956)
H York, *Race to Oblivion* (New York, 1970)
AB Zegart, *Flawed by Design* (Stanford, CA, 1999)
VM Zubok, *Khrushchev and the Berlin Crisis, 1958–1962* (Washington, DC, 1993)

Index

About the Author

Mark Shanahan is a Lecturer in Politics and International Relations at the University of Reading in the UK. He moved into academia after more than 20 years in journalism and business communications and earned his PhD from Brunel University, London. In addition to being Departmental Director of Teaching & Learning, he teaches a number of courses in both British and US Government, and researches in recent US and UK political history, notably on the celebretisation of politics.

Lightning Source UK Ltd.
Milton Keynes UK
UKOW08n2210210417
299653UK00001B/24/P

9 781498 528146